XXXIII Symposium on Nuclear Physics 2010

Journal of Physics: Conference Series Volume 239

Morelos, Mexico
5-8 January 2010

Editors:

Libertad Barron-Palos
Roelof Bijker
Ruben Fossion
David Lizcano

ISBN: 978-1-61738-800-2
ISSN: 1742-6588

Printed from e-media with permission by:

Curran Associates, Inc.
57 Morehouse Lane
Red Hook, NY 12571

Some format issues inherent in the e-media version may also appear in this print version.

Printed by Curran Associates, Inc. (2010)

For permission requests, please contact the Institute of Physics
at the address below.

Institute of Physics
Dirac House, Temple Back
Bristol BS1 6BE UK

Phone: 44 1 17 929 7481
Fax: 44 1 17 920 0979

techtracking@iop.org

Additional copies of this publication are available from:

Curran Associates, Inc.
57 Morehouse Lane
Red Hook, NY 12571 USA
Phone: 845-758-0400
Fax: 845-758-2634
Email: curran@proceedings.com
Web: www.proceedings.com

TABLE OF CONTENTS

012001 Perspectives on 1/f Noise in Quantum Chaos 1
R A Molina, A Relaño, J Retamosa, L Muñoz, E Faleiro, J M Gómez

012002 Chaotic Dynamics in Collective Models of Nuclei 7
P Stránský, M Macek, P Cejnar, A Frank, R Fossion, E Landa

012003 Ultra-high-energy Cosmic Ray and Neutrino Physics Using the Moon 14
Olaf Scholten

012004 Nuclear Reaction Rates and Opacity in Massive Star Evolution Calculations 21
D Bahena, J Klapp, H Dehnen

012005 Phase Transitions in Algebraic Cluster Models 29
H Yépez-Martínez, L Parra Rodríguez, P O Hess, J Cseh, G Lévai

012006 On the Hyperdeformed State of the ^{36}Ar Nucleus 36
J Cseh, J Darai, N V Antonenko, G G Adamian, A Algora, P O Hess, A Lépine-Szily

012007 Single-particle Structure of Neutron-rich Nuclei 43
J A Cizewski, K L Jones, R L Kozub, S D Pain

012008 Proton-neutron Pairing Correlations in the Nuclear Shell Model 47
L Yang, Y-M Zhao, S Pittel, B Thakur, N Sandulescu, A Poves

012009 Structure of the Nucleon in the Unquenched Quark Model 53
Roelof Bijker, Elena Santopinto

012010 Neutral Current Neutrino-nucleon Interaction and the $\Delta(1232)$ 59
A Mariano, C Barbero, G López Castro

012011 Exploring the Nucleon Structure from First Principles of QCD 66
W Bietenholz, N Cundy, M Göckeler, R Horsley, H Perlt, D Pleiter, P E L Rakow, G Schierholz, A Schiller, T Streuer, J M Zanotti

012012 NPDGamma: A Measurement of the Parity-violating Gamma Asymmetry A_γ in N-P Capture 74
R C Gillis, R Alarcon, S Balascuta, L Barrón-Palos, J D Bowman, R D Carlini, W Chen, T E Chupp, C Crawford, S Covrig, M Dabaghyan, N Fomin, S J Freedman, T R Gentile, M T Gericke, G L Greene, F W Hersman, T Ino, G L Jones, B Lauss, M Leuschner, W R Lozowski, R Mahurin, Y Masuda, J Mei, G S Mitchell, S Muto, H Nann, S A Page, S I Penttilä, W D Ramsay, A Salas-Bacci, S Santra, M Sharma, P-N Seo, E I Sharapov, T B Smith, W M Snow, W S Wilburn, V Yuan

012013 A Measurement of Correlation Parameters in the Decay of Polarized Free Neutrons:The abBA Experiment 82
L Barrón-Palos, E Chávez, C Crawford, Q Curiel-García, A Huerta, M A Juárez-Rosete, D J Marín-Lámbarri, E Martin, M E Ortiz, S I Penttilä, P Rodríguez-Zamora, A Salas, Z Tang, W S Wilburn

012014 Studies with Radioactive Ion Beams in the A ~ 80 Region 90
E Padilla-Rodal

012015 Calculations on the Threshold Anomaly of Weakly Bound Projectiles with São Paulo and Woods-Saxon Polarization Potentials 100
A Gómez-Camacho, E F Aguilera, E Martínez-Quiroz, P R S Gomes, J Lubian, L F Canto

Author Index

International Advisory Committee

Richard F. Casten	Yale, USA
Osvaldo Civitarese	La Plata, Argentina
Jerry P. Draayer	LSU, USA
Alfredo Galindo-Uribarri	ORNL & UT, USA
Jozsef Cseh	ATOMKI, Hungary
James J. Kolata	Notre Dame, USA
Luiz Carlos Chamon	São Paulo, Brazil
Jorge López	UTEP, USA
Joseph B. Natowitz	Texas A & M, USA
Ma. Esther Ortiz	IF-UNAM
Stuart Pittel	Delaware, USA
Andrés Sandoval	IF-UNAM
Piet Van Isacker	GANIL, France
Michael Wiescher	Notre Dame, USA

Organizing Commmittee

Libertad Barrón-Palos	IF-UNAM
Roelof Bijker (Chair)	ICN-UNAM
Ruben Fossion	ICN-UNAM
David Lizcano	ININ

Sponsors

Instituto de Ciencias Nucleares, UNAM
Instituto de Física, UNAM
Instituto Nacional de Investigaciones Nucleares
División de Física Nuclear de la SMF
Centro Latinoamericano de Física

Preface

In the present volume of *Journal of Physics: Conference Series* we publish the proceedings of the "XXXIII Symposium on Nuclear Physics", that was held from January 5-8, 2010 at the Hacienda Cocoyoc, Morelos, Mexico. The proceedings contain the plenary talks that were presented during the conference. The abstracts of all contributions, plenary talks and posters, were published in the Conference Handbook.

The Symposium on Nuclear Physics has a long and distinguished history. From the beginning it was intended to be a relatively small meeting designed to bring together some of the leading nuclear scientists in the field. Its most distinctive feature is to provide a forum for specialists in different areas of nuclear physics, both theorists and experimentalists, students, postdocs and senior scientists, in a relaxed and informal environment providing them with a unique opportunity to exchange ideas. After the first meeting in Oaxtepec in 1978, the Symposium was organized every year without interruption which makes the present one the 33rd in a row.

This year's meeting was dedicated to the memory of Marcos Moshinsky, who passed away on April 1, 2009. Dr. Moshinsky was the most distinguished pioneer and promoter of nuclear physics in Mexico and Latin America and holds the record of 31 (out of 32) participations at the Symposium. In the inaugural session, Alejandro Frank (ICN-UNAM), Peter Hess (ICN-UNAM) and Jorge Flores (IF-UNAM) spoke in his honor and recalled the virtues that characterized him as a teacher, scientist, founder of schools and academic institutions, colleague and friend. His generosity, excellence and honesty were emphasized as the personal qualities that characterized both his personal and academic life.

Marcos Moshinsky (1921-2009)

The scientific program consisted of 26 invited talks and 20 posters on a wide variety of hot topics in contemporary nuclear physics, ranging from the traditional fields of nuclear structure and reactions to radioactive beams, nuclear astrophysics, hadronic physics, fundamental symmetries, ultrarelativistic heavy ions, cosmic rays and quantum chaos. The high quality of the talks, the prestige of the speakers and the broad spectrum of subjects covered in the meeting, shows that nuclear physics is a very active area at the frontier of scientific research which establishes bridges between many different disciplines.

One of the exciting new areas in nuclear physics is that of radioactive beams. It provides a powerful tool not only to study exotic nuclei close to the proton and neutron drip lines to obtain important information about the nature of the nucleon-nucleon interaction in stable and unstable nuclei, but also to address questions of fundamental importance in nuclear astrophysics as well as in various applications like mass spectroscopy, the production of radioactive isotopes and medical applications (Galindo-Uribarri). There was a presentation on the FRIB project which is currently under construction in the USA and its relevance for nuclear astrophysics and the limits of stability (Sherrill and Schatz) In the session on nuclear structure, there were several talks on the nucleon-nucleon interaction in nuclei close to the proton and neutron drip lines, like neutron-proton pairing in nuclei with an equal number of protons and neutrons (Pittel), and studies of stable and unstable neutron-rich nuclei near the closed shells $N = 82$ and $N = 50$ using (d, p) transfer reactions and Coulomb excitation by means of radioactive beams (Cizewski and Padilla-Rodal). There were several talks on the importance of reaction rates for the excitation of spin-isospin resonances (Sakai), massive star evolution (Klapp) and nuclear synthesis and stellar evolution (Rolfs). In another presentation, the importance of rare isotopes for astrophysical processes was highlighted (Schatz).

In addition, there were discussions about the importance of the Pauli exclusion principle and phase transitions in nuclear cluster models (Cseh and Hess) and an analysis of elastic scattering close to the Coulomb barrier in the framework of the optical model (Gómez-Camacho).

Another important area is that of hadronic physics, the study of nucleons as composite systems of strongly interacting quarks and gluons. This field is situated on the borderline between nuclear and particle physics and presents a formidable challenge since the scale of the energies involved prohibits the use of the methods of perturbative QCD. In the session on hadronic physics, there were talks on chiral symmetry in non-perturbative QCD (Bietenholz), the structure of the nucleon in an unquenched quark model (Bijker), quark and meson degrees of freedom in Deeply Virtual Compton Scattering (Szczepaniak) and studies of hadronic structure by means of neutrino-induced pion production (Mariano).

At ultrarelativistic energies, one can study the phase transition between hadronic matter and a new state of nuclear matter, the quark-gluon plasma. The Organizing Committee is proud to mention that the Cocoyoc 2010 meeting was one of first international conferences where the first scientific resuls from LHC were presented by the ALICE collaboration on proton-proton collisions at an energy of 900 GeV (Paic).

In the session on cosmic rays, there was a talk on the origin of ultra high-energy cosmic rays (100 EeV $= 10^{20}$ eV) as observed by the Pierre Auger Observatory (Medina-Tanco). Furthermore, there was a presentation of the project NuMoon which proposes to use the Moon's surface to detect cosmic rays that are out of the range of the Auger Observatory (Scholten). In addition, there was a review of the advances of the Mexican project HAWC, an observatory under construction in the Sierra Negra of Puebla (Sandoval) to study some of the most violent phenomena in the Universe through the detection of gamma rays with energies between hundreds of GeV and hundreds of TeV.

At the other extreme of the energy scale is the field of the study of fundamental symmetries with novel experiments with ultra-cold neutrons. In this session, there were proposals by the NPDGamma collaboration (Gillis) and the abBA collaboration (Barrón-Palos) to use polarized neutrons to study the weak hadronic interaction and neutron beta-decay, respectively. In another talk, there was a discussion on the measurement of the electric dipole moment of the neutron which is of fundamental importance to the standard model (Crawford).

Finally, there was a session dedicated to quantum chaos where various proposals were discussed to identify chaotic behaviour in quantum systems, such as spectral fluctuations, time series and $1/f$ noise (Molina). Applications were presented to many-body systems in nuclear physics (Stránský) and the dripping-laser system as a quantum realization of the dripping faucet, one of the paradigms of classical chaos (Fossion).

Many of the participants spent the free afternoon in the "magic village" of Tepoztlán to visit the arts and crafts market, the church, the convent or just to enjoy the good food. The most daring persons climbed the nearby cliff Tepozteco and the pyramid built on top of it to contemplate the spectacular view over the valley. As always, some of the participants of the conference prolonged their stay in Mexico to establish, to develop or to consolidate their collaborations with the local nuclear physics community.

At the annual reunion of the Nuclear Physics Division which was held during the meeting, there was a long discussion on the future of the Symposium and, more generally, of the field of nuclear physics in Mexico and Latin America. Libertad Barrón Palos, Roelof Bijker, Ruben Fossion and David Lizcano were re-elected as members of the Local Organizing Committee of the next Symposium which will be held at the Hacienda Cocoyoc from January 4-7, 2011, but now with Libertad as Chair.

Libertad Barrón-Palos
Roelof Bijker
Ruben Fossion
David Lizcano

XXXIII SYMPOSIUM ON NUCLEAR PHYSICS
JANUARY TUESDAY 5 - FRIDAY 8, 2010
COCOYOC MORELOS, MÈXICO.

Participants

Elí F. Aguilera (ININ, Mexico)
E-mail: eli.aguilera@inin.gob.mx
Armando Barrañón Cedillo (UAM-A, Mexico)
E-mail: bca@correo.azc.uam.mx
Libertad Barrón-Palos (IF-UNAM, Mexico)
E-mail: libertad@fisica.unam.mx
Tatyana Belyaeva (UAEM, Mexico)
E:mail: tbl@uaemex.mx
Wolfgang Bietenholz (ICN-UNAM, Mexico)
E-mail: wolbi@nucleares.unam.mx
Roelof Bijker (ICN-UNAM, Mexico)
E-mail: bijker@nucleares.unam.mx
Jolie Cizewski (Rutgers, USA)
E-mail: cizewski@physics.rutgers.edu
Christopher Crawford (Kentucky, USA)
E-mail: crawford@pa.uky.edu
Jozsef Cseh (ATOMKI, Hungary)
E-mail: cseh@atomki.hu
Judit Darai (ATOMKI, Hungary)
E-mail: darai@atomki.hu
Jorge Flores (IF-UNAM, Mexico)
E-mail: jfv@fisica.unam.mx
Víctor Hugo Flores Soto (IIM-UNAM, Mexico)
E-mail: poeta355@hotmail.com
Ruben Fossion (ICN-UNAM, Mexico)
E-mail: fossion@nucleares.unam.mx
Alejandro Frank (ICN-UNAM, Mexico)
E-mail: frank@nucleares.unam.mx
Paul Fraser (ICN-UNAM, Mexico)
E-mail: paul.fraser@nucleares.unam.mx
Alfredo Galindo-Uribarri (ORNL, USA)
E-mail: uribarri@mail.phy.ornl.gov
Ronald Fernando García Ruiz (ICN-UNAM, Mexico)
E-mail: rf8886@gmail.com
Hugo García Tecocoatzi (ICN-UNAM, Mexico)
E-mail: hugogt_1597@hotmail.com
Chad Gillis (Indiana, USA)
E-mail: rcgillis@indiana.edu
Arturo Gómez-Camacho (ININ, Mexico)
E-mail: arturo.gomez@inin.gob.mx
Peter Hess (ICN-UNAM, Mexico)
E-mail: hess@nucleares.unam.mx
Jorge Hirsch (ICN-UNAM, Mexico)
E-mail: hirsch@nucleares.unam.mx
Arcadio Huerta (IF-UNAM, Mexico)
E-mail: arcadio@fisica.unam.mx

Araceli Ibañez-Sandoval (ESIME-IPN, Mexico)
E-mail: aibanez12@gmail.com
Miguel Ángel Juárez-Rosete (IF-UNAM, Mexico)
E-mail: juaros24@gmail.com
Jaime Klapp (ININ, Mexico)
E-mail: jaime.klapp@inin.gob.mx
Emmanuel Landa (ICN-UNAM, Mexico)
E-mail: emmanuel.landa@nucleares.unam.mx
Sergio Lerma (UV, Mexico)
E-mail: slerma@uv.mx
David Lizcano (ININ, Mexico)
E-mail: david.lizcano@inin.gob.mx
Ruslán Magaña (ICN-UNAM, Mexico)
E-mail: ruslan.magana@nucleares.unam.mx
Marissa Maldonado Velázquez (ININ, Mexico)
E-mail: mmv.marissa@gmail.com
Alejandro Mariano (La Plata, Argentina)
E-mail: mariano@fisica.unlp.edu.ar
Daniel José Marín Lámbarri (IF-UNAM, Mexico)
E-mail: dookye@gmail.com
Jossellin Nayelly Martínez-García (UNAM, Mexico)
E-mail: jossellin@gmail.com
Enrique Martínez-Quiroz (ININ, Mexico)
E-mail: enrique.martinez@inin.gob.mx
Gustavo Medina Tanco (ICN-UNAM, Mexico)
E-mail: gmtanco@nucleares.unam.mx
Joel Mendoza Temis (ICN-UNAM, Mexico)
E-mail: temis@nucleares.unam.mx
Rafael Molina (Madrid, Spain)
E-mail: molina@iem.cfmac.csic.es
Juan Guillermo Mungía Fernández (ICN-UNAM, Mexico)
E-mail: juanguillermo@ciencias.unam.mx
Ma. Esther Ortiz (IF-UNAM, Mexico)
E-mail: ortiz@fisica.unam.mx
Elizabeth Padilla-Rodal (ICN-UNAM, Mexico)
E-mail: padilla@nucleares.unam.mx
Guy Paic (ICN-UNAM, Mexico)
E-mail: guypaic@nucleares.unam.mx
Stuart Pittel (Delaware, USA)
E-mail: pittel@bartol.udel.edu
Jorge Arturo Rojas-Santana (UAEM, Mexico)
E-mail: arturo_santana00@hotmail.com
Claus Rolfs (Bochum, Germany)
E-mail: rolfs@ep3.ruhr-uni-bochum.de
Antonio Ruiz Sánchez (ININ, Mexico)
E-mail: an.ruiz.san@gmail.com
Hideyuki Sakai (Tokyo, Japan)
E-mail: sakai@phys.s.u-tokyo.ac.jp

Andrés Sandoval (IF-UNAM, Mexico)
E-mail: asandoval@fisica.unam.mx
Hendrik Schatz (MSU, Mexico)
E-mail: schatz@nscl.msu.edu
Olaf Scholten (KVI, The Netherlands)
E-mail: scholten@kvi.nl
Leonid Serkin (ICN-UNAM, Mexico)
E-mail: lserkin@nucleares.unam.mx
Pavel Stránský (ICN-UNAM, Mexico)
E-mail: pavel.stransky@nucleares.unam.mx
Adam Szczepaniak (Indiana, USA)
E-mail: aszczepa@indiana.edu
Víctor Velázquez (FC-UNAM, Mexico)
E-mail: vmva@hp.fciencias.unam.mx
Huitzilin Yépez Martínez (UACM, Mexico)
E-mail: huitzilin.yepez@nucleares.unam.mx
Tochtli Cuauhtli Yépez Martínez (ICN-UNAM, Mexico)
E-mail: tochestaz@yahoo.com

Invited Talks

- Rafael Molina (Madrid)
  Applications of $1/f$ noise to spectral statistics
- Ruben Fossion (ICN-UNAM)
  The dripping laser: quantum chaos and $1/f$ noise in a phase transition in light
- Pavel Stránský (ICN-UNAM)
  Chaotic dynamics in collective models of nuclei
- Gustavo Medina-Tanco (ICN-UNAM)
  New developments in the area of ultra-high energy cosmic rays
- Olaf Scholten (KVI)
  High-energy cosmic rays
- Andrés Sandoval (IF-UNAM)
  Title: HAWC, a High Energy Gamma Ray Observatory in Mexico
- Hendrik Schatz (MSU)
  Nuclear astrophysics with rare isotope beams
- Claus Rolfs (Bochum)
  Heavy-ion burning in stars
- Jaime Klapp (ININ)
  On the influence of opacity and nuclear reaction rates on the numerical evolution of massive stars
- Peter Hess (ICN-UNAM)
  Phase transitions in nuclear cluster models
- Jozsef Cseh (ATOMKI)
  Quasi-dynamical SU(3) symmetry and nuclear clusterization
- Jolie Cizewski (Rutgers)
  Single-particle structure of neutron rich nuclei
- Stuart Pittel (Bartol)
  Proton-neutron pairing in the nuclear shell model
- Roelof Bijker (ICN-UNAM)
  Symmetries of the unquenched quark model
- Alejandro Mariano (La Plata)
  Neutral current neutrino-nucleon interaction and the isobar $\Delta(1232)$
- Adam Szczepaniak (Indiana)
  Nucleon structure as seen by Deeply Virtual Compton Scattering
- Wolfgang Bietenholz (ICN-UNAM)
  Chirality and Operator Product Expansion in non-perturbative QCD
- Christopher Crawford (Kentucky)
  The neutron EDM experiment at the SNS
- Chad Gillis (Indiana)
  The NPDGamma experiment
- Libertad Barrón-Palos (IF-UNAM)
  Measurement of the correlation parameterers in the decay of free polarized neutrons: the abBA experiment
- Guy Paic (ICN-UNAM)
  The first ALICE results with pp collisions at 900 GeV

- Brad Sherrill (MSU)
  Search for the origins and limits of nuclei using the Facility for Rare Isotope Beams
- Elizabeth Padilla-Rodal (ICN-UNAM)
  The use of radioactive ion beams in the $A \sim 80$ *region*
- Alfredo Galindo-Uribarri (ORNL)
  Applications with unstable nuclei
- Arturo Gómez-Camacho (ININ)
  The threshold anomaly of weakly bound projectiles from recent elastic scattering measurements around the Coulomb barrier
- Hideyuki Sakai (Tokyo)
  Exothermic heavy-ion charge-exchange reactions with unstable beams and the high resolution magnetic spectrometer SHARAQ

Posters

- Armando Barrañón (UAM-A)
  Non-monoticity of scaled invariance in heavy-ion collisions via Latino model
- Tatyana Belyaeva (UAEM), E.F. Aguilera (ININ), E. Martínez-Quiroz (ININ), A.M. Moro (Sevilla, Spain) and J.J. Kolata (Notre Dame, USA)
  Astrophysical S_{17} factor extraction from breakup of ^{8}B on ^{58}Ni at energies around the Coulomb barrier
- Judit Darai (ATOMKI)
  Elongated shape isomers and clusterization of atomic nuclei
- Paul Fraser (ICN-UNAM), K. Amos (Melbourne, Australia), L. Canton, G. Pisent (Padova, Italy), S. Karataglidis (Rhodes, South Africa), J.P. Svenne (Manitoba, Winnipeg), D. van der Kniff (Melbourne, Australia)
  Coupled-channel evaluations of cross sections for scattering involving particle-unstable resonances
- Hugo García Tecocoatzi (ICN-UNAM)
  Systematics of nuclear masses and the shell model
- Araceli Ibañez-Sandoval (ESIME-IPN), V. Velázquez (FC-UNAM), M.E. Ortiz (IF-UNAM), A. Frank (ICN-UNAM) and C. Vargas (UV)
  Quantum chaos evolution from shell model calculations
- Miguel Ángel Juárez-Rosete (IF-UNAM) *Precision neutron polarimetry*
- Emmanuel Landa, R. Fossion, P. Stránský, I. Morales, J.C. López-Vieyra, A. Frank (ICN-UNAM) and V. Velázquez (FC-UNAM)
  Scale invariance in chaotic time series: classical and quantum examples
- Sergio Lerma (UV)
  The wave function of the isovectorial and isoscalar pairing Hamiltonian from a Bethe-Ansatz approach
- David Lizcano (ININ)
  Alpha-particles from $^{6}Li + ^{58}Ni$
- Ruslán Idelfonso Magaña Vsevolodovna (ICN-UNAM)
  Correlations between nucleon transfer reactions in nuclear supersymmetry
- Marissa Maldonado Velázquez (ININ)
  Resonant elastic scattering of $^{12}C + ^{4}He$
- Daniel José Marín Lámbarri, Libertad Barrn Palos, Quiela Marina Curiel Garca, Efraín Rafael Chávez Lomelí, Arcadio Huerta Hernández, Miguel Ángel Juárez Rosete, María Esther Ortiz Salazar, Penélope Rodríguez Zamora, José Francisco Favela Pérez (IF-UNAM), Ghiraldo Murillo Olayo, Rafael Policroniades Rueda, Armando Varela González (ININ)
  Development of an adiabatic spin flipper for slow neutrons
- Jossellin Nayelly Martínez-García (UNAM)
  Development of radioactive beams using a Tandem Accelerator
- Enrique Martínez-Quiroz, E.F. Aguilera, A. Gómez-Camacho, D. Lizcano, P. Rosales, A. Ruiz-Sánchez (ININ), J.J. Kolata (Notre Dame, USA)
  Fusion measurements through evaporated particcles
- Joel Mendoza Temis (ICN-UNAM)
  The anatomy of the simplest Duflo-Zuker mass formula
- Jorge Arturo Rojas-Santana and T.L. Belyaeva (UAEM) *Schrödinger soliton scattering on the linear potential: exactly integrable models*

- Antonio Ruiz Sánchez (ININ, FC-UAEM)
  Optical model analysis of the system $^{7}Be + {}^{58}Ni$ *at energies around the Coulomb barrier*
- Leonid Serkin (ICN-UNAM)
  Jet Physics at ALICE
- Tochtli Cuauhtli Yépez Martínez (ICN-UNAM)
  A solvable model for many quarks systems in QCD Hamiltonains

All papers published in this volume of *Journal of Physics: Conference Series* have been peer reviewed through processes administered by the proceedings Editors. Reviews were conducted by expert referees to the professional and scientific standards expected of a proceedings journal published by IOP Publishing.

Perspectives on $1/f$ noise in quantum chaos

R A Molina[1], A Relaño[2], J Retamosa[2], L Muñoz[2,3], E Faleiro[3] and J M G Gómez[2]

[1]Instituto de Estructura de la Materia - CSIC, Serrano, 123, 28006 Madrid, Spain.
[2] Departamento de Física Atómica, Molecular y Nuclear, Universidad Complutense de Madrid, E-28040 Madrid, Spain.
[3] Departamento de Física Aplicada, E. U. I. T. Industrial, Universidad Politécnica de Madrid, E-28012 Madrid, Spain.

E-mail: `molina@iem.cfmac.csic.es`

Abstract. The power spectrum of the δ_n statistic of quantum spectra presents $1/f^\alpha$ noise. For chaotic systems $\alpha = 1$ while for regular systems $\alpha = 2$. Although the transition from regularity to chaos is non universal, for a wide variety of systems with a mixed phase space the value of α is intermediate between 1 and 2 and can be related to the fraction of regular or chaotic orbits in the total phase space. This statistic can be a very useful tool for the analysis of experimental spectra, specially in the case of missing levels or spectral sequences with mixed symmetries.

1. Introduction

The study of energy level fluctuations is a basic tool in understanding quantum chaos. The pioneering work of Berry and Tabor showed that the spectral fluctuations of a quantum system whose classical analogue is fully integrable are well described by Poisson statistics [1] meaning that the energy levels behave as uncorrelated random variables. On the other hand, Bohigas *et al.* conjectured that the fluctuation properties of quantum systems which in the classical limit are fully chaotic coincide with those of Random Matrix Theory (RMT) [2] showing strong level repulsion and long-range correlations. The Bohigas-Giannoni-Schmit conjecture has been explained in the context of semiclassical periodic-orbit theory [3].

In the last decade a new approach to quantum chaos and spectral fluctuations has been proposed [4]. The main idea behind this approach is to consider the sequence of energy levels as a discrete time series and study level correlations using the tools of time series analysis. The analogy between the energy spectrum and a discrete time series is established in terms of the δ_n statistic, defined as the deviation of the $(n+1)$th level from its mean value. In terms of unfolded energy levels

$$\delta_n = \sum_{i=1}^{n} (s_i - \langle s \rangle) = \epsilon_{n+1} - \epsilon_1 - n, \tag{1}$$

where $s_i = \epsilon_{i+1} - \epsilon_i$ and ϵ_i is the i-th unfolded level. The unfolded energy levels are defined using the average accumulated level density $\overline{N}(E)$ as $\epsilon_i = \overline{N}(E_i)$. This mapping is needed to remove the main trend defined by the smooth part of the level density in order to compare the statistical properties of the spectral fluctuations of different systems or different parts of the

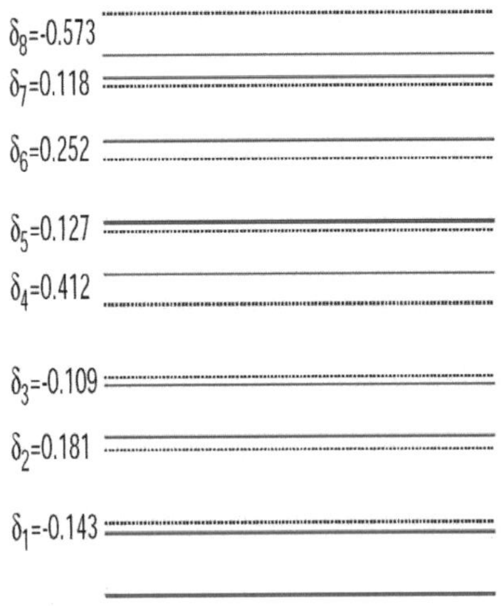

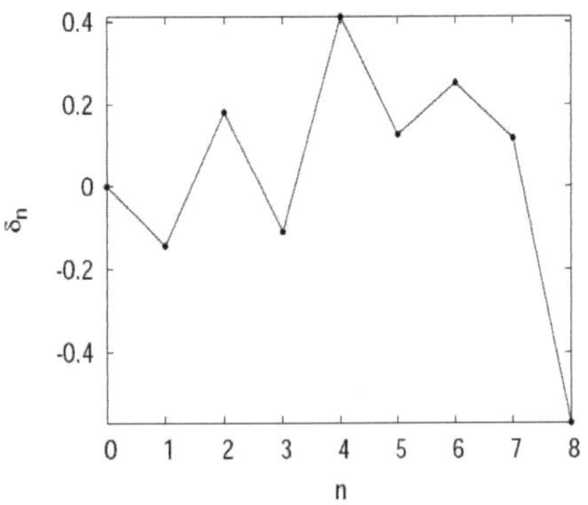

Figure 1. Example of how to build the statistic δ_n from the unfolded spectra. In the top panel we show an spectrum with the unfolded energies ordered in the y-axis. The red full lines is the real spectrum while the blue broken lines mark an hypothetical evenly distributed spectrum. The difference between the position of the red and blue lines is the δ_n showed in the left panel as a function of the level position n. Taking this level ordering n as a pseudotime the δ_n positions can be considered as the time series of the position of a Brownian particle moving in one-dimension.

same spectrum. In the language of time series analysis the unfolding mapping is a procedure for making stationary the discrete time series defined by the δ_n, its average and fluctuations not depending on time. In Fig. 1 we show an example of the procedure of building the δ_n statistic from the sequence of unfolded levels.

There are some differences between the δ_n statistic and the displacement of a Brownian particle. For example, $s_k - 1 > -1$ as there are no negative spacings by definition. Also the displacement of a Brownian particle usually depends on the past trajectory while the position of the energy levels and consequently the value of $s_k - 1$ depends not only on the lower energy levels, but also on the upper levels so considered as a time series there is a dependence on the past as well as on the future. Nevertheless, in spite of these peculiarities, the analogy exists, and the function δ_n is the analogue of a Brownian particle displacement in one-dimension.

Once we make the interpretation of the δ_n as a discrete time series we can use the powerful tools of time series analysis to get information about the dynamics of the quantum system. One of the most important and ubiquitous properties of time series is self-similarity, defined as the invariance of the signal $X(t)$ under a global scale transformation $X(\lambda t) \propto X(t)$. In real physical systems there are always limiting time scales limiting strict self-similarity so this property only appears in a finite range of time scales. To unveil the characteristic time scales of a signal $X(t)$ the Fourier transform can be used. The power spectrum $P(\omega)$ is normally used to avoid working with complex systems. Self-similarity in the frequency space can also appear.

Although the reasons are poorly understood, time series describing many different natural and social phenomena present a power spectrum behaving as

$$P(\omega) \propto \omega^{-\alpha}, \tag{2}$$

where $\alpha \geq 1$ [5, 6]. The signals of this type, generically known as presenting $1/f^{\alpha}$ noise, have no characteristic time scale, and their correlation time is comparable to the duration of the whole time series. Surprisingly enough in many different systems the power law is maintained over many orders of magnitude and the term "ubiquitous $1/f$ noise" was coined to describe this fact.

2. Theoretical derivation

The power spectrum of the δ_n can be calculated by

$$P_k^\delta = \frac{1}{N} \sum_{p,q}^{N-1} \delta_p \delta_q \exp\left(-\frac{2\pi i k(p-q)}{N}\right), \tag{3}$$

where N is the number of levels in the sequence. In the first work that published results on this quantity [4] it was numerically found that regular systems present $1/f^2$ noise while chaotic systems present $1/f$ noise and this difference was proposed to be a general characterization of the spectral statistics of quantum chaotic systems. Numerical results from the Two Body Random Ensemble were shown also to present $1/f$ noise [7]. Two years later a theoretical derivation of $1/f$ and $1/f^2$ noises in chaotic and regular systems from RMT was reported by our group [8]. The power spectrum of the δ_n statistic was expressed in terms of the average power spectrum of the fluctuating part of the level density, known as spectral form factor $K(\tau)$.

$$\begin{aligned} \left\langle P_k^\delta \right\rangle &= \frac{N^2}{4\pi^2}\left[\frac{K(k/N)-1}{k^2} + \frac{K(1-k/N)-1}{(N-k)^2}\right] \\ &+ \frac{1}{4\sin^2(\pi k/N)} + \Delta, \quad \text{for } N \gg 1, \end{aligned} \tag{4}$$

where $k = 1, 2, ..., N-1$ and $\Delta = -1/12$ for large energy level sequences $N \gg 1$ belonging to the three classical RMT and their interpolations while $\Delta = 0$ for Poisson spectra [8, 9]. Exact analytical expressions for the spectral form factor are known for the Gaussian Ensembles (GE) of RMT and for the Poisson ensemble. The important behavior for our purpose is that $K(\tau) \propto \tau$ for $\tau \ll 1$ for the GE while is constant for the Poisson ensemble. Substituting the values of the spectral form factor for the different ensembles we can obtain the final result for the GE in the small frequency region

$$\left\langle P_k^\delta \right\rangle \sim \frac{N}{2\nu\pi^2 k}, \quad 1 \leq k \ll N, \quad \nu = 1, 2. \tag{5}$$

where $\nu = 1$ corresponds to the GOE ensemble and $\nu = 2$ to the GUE case. Similarly, for integrable systems and $k \ll d$, Eq. (4) becomes

$$\left\langle P_k^\delta \right\rangle \sim \frac{N^2}{4\pi^2 k^2}, \quad 1 \leq k \ll N. \tag{6}$$

These expressions show that for small values of k (small frequencies), the fluctuations of the excitation energy exhibit $1/f$ noise in chaotic systems and $1/f^2$ noise in integrable systems. As the frequency approaches the Nyquist frequency [10] $k = d/2$, higher order terms start becoming really appreciable.

The previous formulas describe not only the $1/f$ or $1/f^2$ behavior correctly, characteristic of small and intermediate frequencies, but also deviations at high frequency due to the Δ parameter. Also the factor preceding the power law is perfectly accounted for. An important point worth mentioning is that RMT is one of the few examples where $1/f^{\alpha}$ noise behavior have been theoretically derived. Experimentally, the presence of $1/f$ noise in microwave billiards was demonstrated in [11]. The theoretical expressions as a function of $K(\tau)$ paved the way to further applications of the power spectrum of the δ_n and of the implications of considering the level sequence as a time series.

3. Applications to imperfect spectra

The statistical analysis of spectral fluctuations and the comparison with RMT results requires energy level sequences to be pure, i. e., all the states of a sequence must have the same quantum numbers (J^π or $J^\pi T$), and there must be no missing or spurious levels in any sequence. Experimentally, it is very difficult to guarantee that these conditions are fulfiled. Missing or spurious levels and mixed symmetries would induce misleading results about the chaotic or regular nature of the quantum systems involved. It is possible to generalize Eq. (4) to the general case of l different symmetries mixed in the sequence, each with a fraction of total number of N levels ϕ_i and each with a fraction of observed levels f_i

$$\begin{aligned} \left\langle P_k^\delta \right\rangle &= \frac{N^2}{4\pi^2}\sum_{i=1}^{l}\phi_i f_i \left[\frac{K_i\left(\frac{f_i k}{\phi_i N}\right)-1}{k^2} + \frac{K_i\left(\frac{f_i(N-k)}{d\phi_i}\right)-1}{(N-k)^2}\right] \\ &+ \frac{1}{4\sin^2(\pi k/N)} + \left\langle f^2\right\rangle \Delta, \end{aligned} \tag{7}$$

where $K_i(\tau)$ is the original spectral form factor of the i pure level sequence mixed in the total one that is being analyzed, and $\left\langle f^2\right\rangle = \sum_{i=1}^{l}\phi_i f_i^2$. It is quite clear that in the limit of very large number of mixed symmetries and/or very few number of observed levels the Poisson result is recovered. In order to use Eq. (7) in practice it is unavoidable to assume a form of the function $K_i(\tau)$. Assuming chaotic RMT result is the most useful case. Our group showed that this expression can be useful to estimate the number of missing levels and mixed symmetries in realistic situations [12]. One way to check this assumption could be to estimate independently the fraction of missing levels from the short range correlations by the $P(s)$ [13, 14] and from long range correlations by the power spectrum. If both results are within statistical errors one could be more certain of the plausability of the original assumption.

4. Behavior in mixed systems

One of the most important open problems in the statistical study of quantum spectra is the behavior of the fluctuations in the spectra of mixed systems, systems that are in between integrability and chaos. There is a consensus in the quantum chaos community that this transition is not universal and different systems present different intermediate statistics. As far as the long range correlations are concerned there appear to be two main ways for the transition to develop as a function of a control parameter. In one case, the long range correlations present an intermediate behavior at all scales. Depending on the value of the control parameter the intermediate behavior can be closer to integrability or to chaos. In the other case, there are chaotic correlations but only between levels separated less than some energy difference. For larger energy separations correlations disappear as in the integrable case [15]. The transition parameter controls the value of the correlation energy scale in the spectrum that can be assimilated to the Thouless energy in weakly disordered mesoscopic systems. Of course, systems with all kinds of intermediate behavior between these two extreme cases are possible and found in the literature, complicating the analysis of the spectral statistics of mixed systems.

There have been two very interesting works calculating the power spectrum of semiclassical systems that transit between integrability and chaos as a function of a control parameter. For the systems analyzed in the intermediate regime $1/f^\alpha$ noise was found in the power spectrum of the δ_n statistic with $1<\alpha<2$ [16, 17]. From the theoretical formulas for the power spectrum, Eq. (4), it is easy to see that $1/f^\alpha$ implies a similar fractional exponent in the spectral form factor $K(\tau)\propto\tau^\beta$ with $\beta=2-\alpha$, for $\tau\ll 1$. This behavior is quite difficult to justify theoretically. Semiclassical models predict a mixture of $1/f$ and $1/f^2$ behaviors [18] as RMT models also do [15, 19]. Resorting to a RMT model taking into account Chaos Assisted Tunneling (CAT)[20],

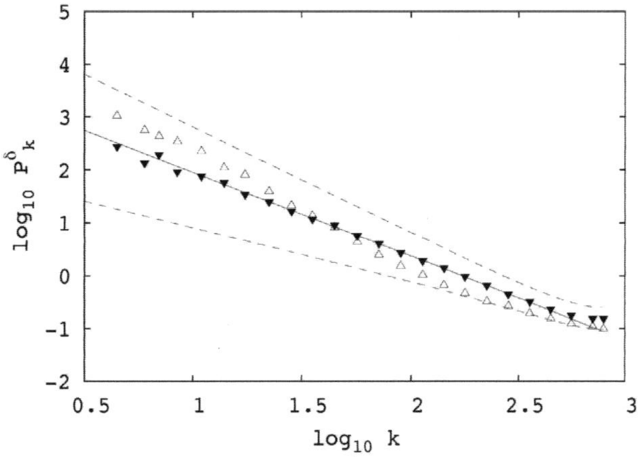

Figure 2. P_k^δ of the RMT model for a mixed system with CAT with an exponential dependence in the energy difference of the variance of the tunneling elements (full triangles) and a constant variance (empty triangles). Comparison with a power law $P_k^\delta = A/k^\alpha$ with $\alpha = 1.58$ (full lines) and the theoretical results for GOE and GDE (broken lines). To obtain the points in the figure a logarithmic binning average was used to reduce the number of points and the fluctuations. 20 matrices of size $N = 3000$ were used for the ensemble average.

recently one of us, A. Relaño, put forward an explanation of the fractional exponent found in mixed systems [21].

We define a RMT model for a mixed system with CAT as

$$H = \begin{pmatrix} GOE & V \\ V & GDE \end{pmatrix}, \tag{8}$$

where GOE and Gaussian Diagonal Ensemble (GDE) represent square diagonal submatrices, whose elements are the eigenvalues of a matrix belonging to the GOE or to GDE, defined as random uncorrelated diagonal matrix. They represent the chaotic energy levels and the regular energy levels of the system, respectively. V determines the tunneling rate between these states. To model CAT in a physically sound way the V matrix is composed by independent Gaussian random variables but the variance is non-constant and depends exponentially in the difference in energies between the regular and the chaotic state connected.

In Fig. 2 we show results for the power spectrum of model of Eq. (8) in the case that the variance of the random elements of V depends exponentially on the difference of energies (full triangles)

$$\sigma_{ij} = \exp\left(-\lambda \left|\bar{E}_i^C - \bar{E}_j^R\right|\right), \tag{9}$$

where λ is the control parameter ($\lambda = 30$ in the figure), $\bar{E}_i^C$ and $\bar{E}_j^R$ are the average position of the $i(j)$ energy levels in the chaotic(regular) region respectively. The average position is used instead of the actual position of the energy levels in order to use the same distribution of the matrix elements of V for the whole ensemble of random matrices instead of a matrix-dependent distribution. The full line in Fig. 2 is a linear fit to the power spectrum that results in slope $\alpha = 1.58 \pm 0.01$. Depending on the value of λ one can obtain different values of the exponent α going from 2 to 1. The empty triangles in the figure are results for the same model of Fig. (2) but with a constant value of $\sigma = 2\dot{1}0^{-3}$. Comparing to the results for Poisson and GOE (broken lines) one can see that the results go from GOE in large values of k to close to Poisson in small values of k.

5. Perspectives

In spite of the great success of the RMT with energy-dependent CAT to explain fractional exponents in the power spectrum that, for example, have been calculated in billiards or perturbed oscillators with a mixed classical phase space [16, 17] there are still many opened questions before fully explaining the behavior of the spectral fluctuations in mixed systems. The first question to answer would be if the modeled form of the matrix elements corresponds to reality in models where $1/f^\alpha$ behavior appears and if the difference in the energy dependence of the variance of the matrix elements is behind the two kinds of behavior of the intermediate systems as it appears from the RMT model simulations.

As mentioned previously, RMT is one of the few cases where the origin of $1/f$ noise has been theoretically explained in detail. Could some kind of RMT be behind the origin of $1/f$ noise in other complex systems besides quantum chaotic ones? We believe more work is needed in this direction.

Acknowledgments

This research has been supported in part by Spanish Government grant No. FIS2009-07277 and No. FIS2006-12783-C03-02, CSPD-2007-00042-Ingenio2010, and by the Universidad Complutense de Madrid grant UCM-910059. A.R. is supported by the CPAN-Consolider program.

References

[1] Berry M V and Tabor M 1977 *Proc. R. Soc. London* A **356** 375
[2] Bohigas O, Giannoni M J and Schmidt C 1984 *Phys. Rev. Lett.* **52** 1
[3] Heusler S, Müller S, Altland A, Braun P and Haake F 2007 *Phys. Rev. Lett.* **98** 044103
[4] Relaño A, Gómez J M G, Molina R A, Retamosa J and Faleiro E 2002 *Phys. Rev. Lett.* **89** 244102
[5] Mandelbrot B B 1999 Multifractals and $1/f$ noise *Springer* New York
[6] Greis N P and Greenside H S 1991 *Phys. Rev.* A **44** 2730
[7] Relaño A, Molina R A and Retamosa J 2004 *Phys. Rev.* E **70** 017201
[8] Faleiro E, Gómez J M G, Mu noz L, Molina R A, Rela no A and Retamosa J 2004 *Phys. Rev. Lett.* **93** 244101
[9] French J B, Kota V K B, Pandey A and Tomsovic S 1988 *Ann. Phys. (N.Y.)* **181** 198
[10] Smith J O 2003 Mathematics of the Discrete Fourier Transform (DFT) ISBN 0-9745607-0-7 [http://ccrma-www.standford.edu/jos/mdft/]
[11] Faleiro E, Kuhl U, Molina R A, Muñoz L, Relaño A and Retamosa J 2006 *Phys. Lett.* A **358** 251
[12] Molina R A and Muñoz L *et al* 2007 *Phys. Lett.* B **644** 25
[13] Agvaanluvsan U, Mitchell G E, Shriner Jr. J F and Pato M P 2003 *Nucl. Instr. Meth.* A **498** 459
[14] Bohigas O and Pato M P 2006 *Phys. Rev.* E **74** 036212
[15] García-García A M 2006 *Phys. Rev.* E **73**, 026213
[16] Gómez J M G, Relaño A, Retamosa J, Faleiro E, Salasnich L, Vranicar M and Robnik M 2005 *Phys. Rev. Lett.* **94** 114101
[17] Santhanam M S and Bandyopadhayay J N 2005 *Phys. Rev. Lett.* **95** 114101
[18] Robnik M 2006 *Int. J. Bifurcation Chaos Appl. Sci. Eng.* **16** 1849
[19] Male C, Le Caër G and Delannay D 2007 *Phys. Rev.* E **76** 042101
[20] Tomsovic S, Ullmo D 1994 *Phys. Rev.* E **50** 145
[21] Relaño A 2008 *Phys. Rev. Lett.* **100** 224101

Journal of Physics: Conference Series **239** (2010) 012002 doi:10.1088/1742-6596/239/1/012002

Chaotic dynamics in collective models of nuclei

P Stránský[1,2], M Macek[1], P Cejnar[1], A Frank[2], R Fossion[2] and E Landa[2]

[1] Institute of Particle and Nuclear Physics, Faculty of Mathematics and Physics, Charles University, V Holešovičkách 2, 180 00 Prague, Czech Republic

[2] Instituto de Ciencias Nucleares, Universidad Nacional Autónoma de México, Apartado Postal 70-543, 04510 México, D.F., Mexico

E-mail: `pavel.stransky@nucleares.unam.mx`

Abstract. We present results of an extensive analysis of classical and quantum signatures of chaos in the geometric collective model (GCM) and the interacting boson model (IBM) of nuclei. Apart from comparing the regular fraction of the classical phase space and the Brody parameter for the nearest neighbor spacing distribution in the quantum case, we also adopt (i) the Peres lattices allowing one to distinguish ordered and disordered parts of spectra and to reveal main ordering principles of quantum states, (ii) the geometrical method to determine the position where the transition from order to chaos occurs, and (iii) we look for the $1/f^{\alpha}$ power law in the power spectrum of energy level fluctuations. The Peres method demonstrates the adiabatic separation of collective rotations in the IBM.

1. Introduction

It has been shown in recent years that simple models of nuclear collective dynamics—the geometric collective model (GCM) and the interacting boson model (IBM)—exhibit a high degree of variability in regular and chaotic features with energy and control parameters [1, 2, 3]. The complex dynamics encoded in relatively simple Hamiltonians well position these systems to be laboratories for detailed investigation into classical-quantum correspondence and for testing different approaches of measuring and visualizing chaos. In addition, it appears that the study of chaos can help in understanding the dynamical structure of the systems and allows for observation of new phenomena, such as the quantum phase transitions [4] and the quasi-dynamical symmetry (QDS) [5].

Classical chaos is commonly studied by means of the visual method of Poincaré sections and numerically characterized by quantities derived from Lyapunov exponents. This approach requires the tedious solution of differential equations of motion for a large set of trajectories in order to obtain an overall image of the degree of regularity. To bypass these difficulties, Horwitz *et al* [6] have proposed recently another way to distinguish between regular and chaotic dynamics. When the system, evolving in a flat Cartesian space with some potential, is transformed to be in a free motion on a special curved manifold, the Riemannian geometry gives a very simple condition of stability: a negative eigenvalue of a matrix obtained from the curvature tensor appearing inside the kinematically accessible area of the system implies instability, and thus chaoticity of the motion.

Quantum measures of chaos are based on certain statistical properties of the energy levels, for example on the nearest-neighbor-spacing distribution (and the related Brody parameter ω) [7].

It is clear that the statistics describe only some bulk features of a sufficiently large portion of the spectrum. The method introduced by Peres [8], however, makes it possible to assign regular, chaotic, or a mixed type of dynamics to *individual states*. It uses lattices formed from the expectation values $A_i = \langle\psi_i|A|\psi_i\rangle$ of an arbitrary operator A plotted against the energies $E_i = \langle\psi_i|H|\psi_i\rangle$. Due to arguments based on the semiclassical EBK quantization, the lattices of points (E_i, A_i) show regular patterns in integrable systems, whereas in the chaotic case the resulting images are formed by a combination of ordered and disordered patterns, whose relative size depends on the degree of chaoticity. Note that the Peres method is entirely visual and can be considered as a quantum analogue of the classical Poincaré sections.

Long-range correlations of eigenlevels are usually studied by the spectral rigidity Δ_3 or the number variance Σ^2, from which, however, it is difficult to quantify the degree of regularity of a system. Incorporating long-range correlations into quantitative studies has been done by Relaño *et al* [9] considering spectral fluctuation of a time series of deviations between the unfolded eigenenergies and uniformly distributed levels. The ensemble average of the power spectrum obeys the power law $\langle S(f)\rangle = 1/f^{\alpha}$ where α, bounded between 1 and 2 (for a totally chaotic and an integrable system, respectively), can serve as a measure of regularity.

The aim of this contribution is to establish the standard methods of classical and quantum chaos in the GCM and IBM and then make a step beyond; we use Peres lattices to present the manifestation of QDS by well-pronounced rotational bands for *high*-energy states in the IBM with symmetry breaking interaction switched on.

2. Geometric collective model

We use the GCM in the nonrotating regime, i.e. considering only the vibrations with zero angular momentum $\boldsymbol{J}$. The Hamiltonian $H = T + V$ describes quadrupole motions of nuclei. Quantizing the system and separating the vibrational and rotational degrees of freedom by moving into the intrinsic frame we obtain the kinetic term in the form

$$T^{5\mathrm{D}} = -\frac{\hbar^2}{2K}\left(\frac{1}{\beta^4}\frac{\partial}{\partial\beta}\beta^4\frac{\partial}{\partial\beta} + \frac{1}{\beta^2\sin 3\gamma}\frac{\partial}{\partial\gamma}\sin 3\gamma\frac{\partial}{\partial\gamma}\right) + T_{\mathrm{rot}}. \tag{1}$$

The system has in total 5 degrees of freedom which is indicated by the label 5D. K is a mass parameter and (β, γ) are the well-known Bohr (shape) coordinates. For states with $J = 0$ the rotational part T_{rot} vanishes. The potential (with adjustable parameters A, B, C) reads as

$$V = A\beta^2 + B\beta^3\cos 3\gamma + C\beta^4. \tag{2}$$

There is another way how to deal with the kinetic term in the case of the principal frame at rest. If the 3 rotational degrees of freedom are frozen *before* the quantization is carried out, we get the familiar 2D kinetic term in polar coordinates (β, γ):

$$T^{2\mathrm{D}} = -\frac{\hbar^2}{2K}\left(\frac{1}{\beta}\frac{\partial}{\partial\beta}\beta\frac{\partial}{\partial\beta} + \frac{1}{\beta^2}\frac{\partial^2}{\partial\gamma^2}\right). \tag{3}$$

Both expressions (1) and (3) have the same classical limit and offer a possibility to study the influence of the quantization method on the quantum chaotic features of the spectrum. Let us emphasize that only the 5D case represents the standard (nuclear) GCM.

We diagonalize both types of the GCM Hamiltonian in the respective 2D or 5D oscillator bases. Due to scaling properties, only one of the parameters (A, B, C, K) in the classical case, or two in the quantum case, determine the qualitative features of the system. We fix $C = 1$ and move along the path of three connected lines (each described by only one principal control parameter), passing all possible configurations of the system: (i) $B \in (0, 1)$ with $A = -1$,

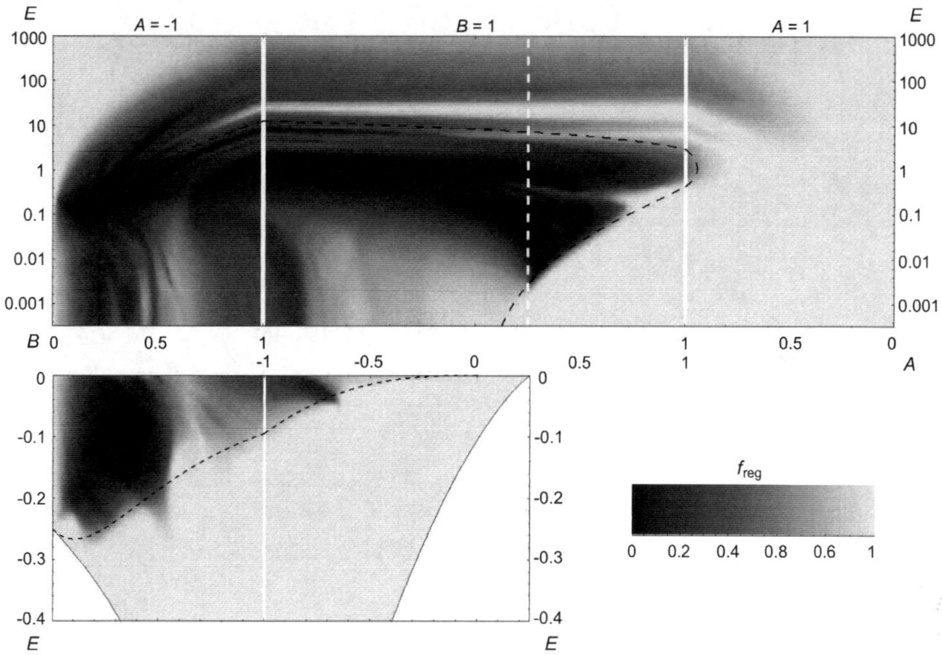

Figure 1. Complete map of the classical regular fraction $f_{\rm reg}$ as a function of control parameters and energy E. The degree of chaos is coded in shades of gray, with light gray (black) corresponding to complete order (chaos). The plot is divided by the thick white lines into three regions of different scaling ($A = -1$, $B = 1$, and $A = 1$, respectively) with free parameters being varied along the horizontal axis. The white dashed line indicates the deformed-spherical shape phase transition. The black dashed line corresponds to the convex-concave transition of the border of kinematically accessible area.

(ii) $A \in (-1, 1)$ with $B = 1$ where the deformed-spherical ground-state shape phase transition is crossed at $(A, B) = (1/4, 1)$, and (iii) $B \in (1, 0)$ with $A = 1$. In the quantum case we consider the classicality constant $\kappa = \hbar/K$ as the second fundamental parameter, which influences only the density of energy levels [10]. We choose two independent Peres operators identified with (i) the square of the angular momentum operator L^2 connected with rotations about angle γ, which is the $J = 0$ projected O(5) invariant (the seniority operator) in the 5D case or the O(2) invariant in the 2D case, and (ii) the nonintegrable perturbation $H' = \beta^3 \cos 3\gamma$ (the Hamiltonian is integrable for $B = 0$ with L^2 being the second integral of motion).

Figure 1 presents the complete map of the classical regularity quantified by the regular fraction of the classical phase space $f_{\rm reg}$. Leaving the leftmost and the rightmost $B = 0$ integrable limits, we observe a progressive retreat of regularity with increase of the perturbation B. In the intermediate domain the structure is extremely complex, forming sharp branching valleys and curved edges. At $B \approx 0.6$ the regularity achieves a well-pronounced maximum, which has been shown to be tightly connected with the IBM semiregular arc [11]. The shape-phase transition region (the white vertical dashed line) is characterized by accentuated chaoticity. The dashed black line encircles the region where the kinematically accessible region in the $(x \equiv \beta \cos\gamma, y \equiv \beta \sin\gamma)$ plane has concave shape. The lower part of the line corresponds exactly with the curve obtained by the Horwitz's geometrical method, which should separate regions with regular and chaotic dynamics. However, it is evident from the figure that this criterion provides only an estimation.

The decay of regular pattern in the quantum Peres lattices when introducing nonintegrability is demonstrated in Figure 2. Starting from a totally ordered case (panel a), small nonintegrable

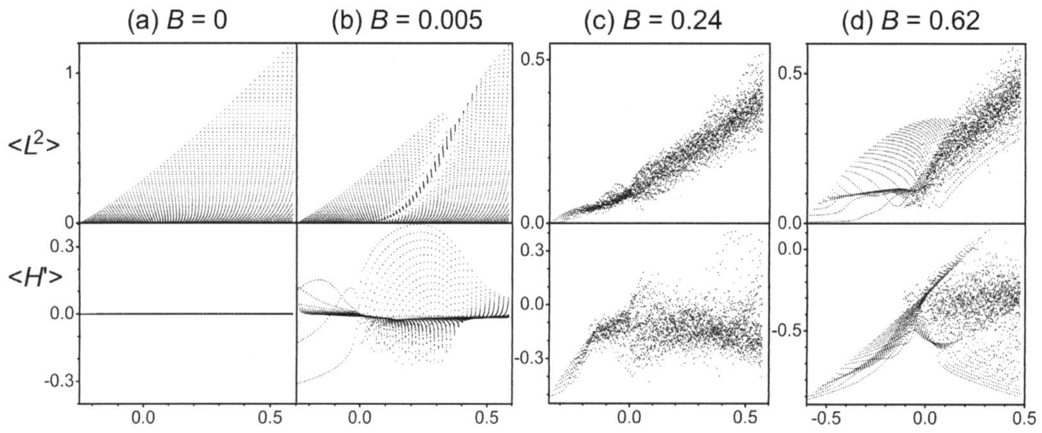

Figure 2. Peres lattices of the 2D GCM for $\kappa = 2.5 \cdot 10^{-5}$ and $A = -1$ with 3000 levels plotted. The transition between integrable (panel a) and nonintegrable (panels b–d) regimes are observed.

perturbation causes disturbances only in a part of a lattice (panel b). Maximal disorder is reached at $B = 0.24$. For higher B the progression reverses and new regular structures start to appear, which is observed in both $\langle L^2 \rangle$ and $\langle H' \rangle$ lattices. The regular islands can be identified with remnants of classical tori.

Figure 3 (a) presents the power spectrum for the method of the $1/f^{\alpha}$ noise in an integrable case of the model. The value of the exponent $\alpha = 2$ is in full agreement with the prediction of the theory [9].

In the quantum case we use two measures of regularity: (i) the adjunct $1 - \omega$ of the Brody parameter, which is estimated by the χ^2 fit applied to the nearest-neighbor spacing distribution of the unfolded levels, and (ii) the quantity $\alpha - 1$ of the $1/f^{\alpha}$ noise in the averaged power spectrum of energy level fluctuation. In order to obtain the energetic dependence of the regularity, we divide the calculated spectra (usually 30 000 well-converging levels) into windows of 1000 levels. Both quantum measures can then be directly compared with f_{reg} [see Figure 3(b)]. The

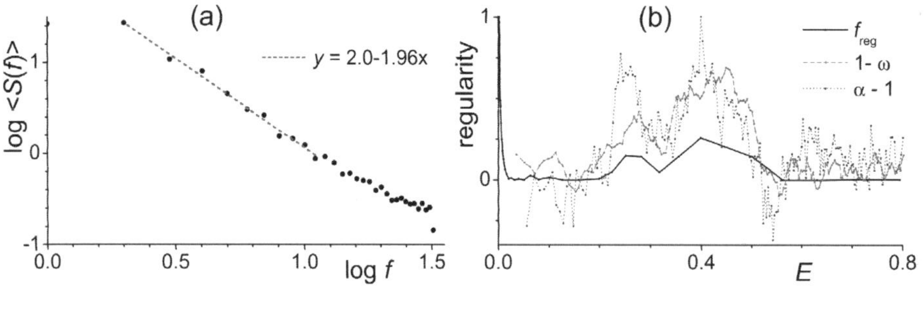

Figure 3. (a) Average power spectrum of the $1/f^{\alpha}$ noise method in the integrable γ-soft case ($A = -1, B = 0$) of the 2D GCM with $\kappa = 4 \cdot 10^{-6}$, using 500 sets of 64 levels, starting from level 8000. We use the simplified notation $x \equiv \log f$ and $y \equiv \log\langle S(f) \rangle$. (b) Comparison of the classical f_{reg} and quantum measures of regularity $1 - \omega$ and $\alpha - 1$ for the 2D GCM in the mixed dynamics configuration with $A = 0.25, B = 1$ and $\kappa = 1 \cdot 10^{-6}$. The standard errors of the measures are 3%, 5% and 20%, respectively.

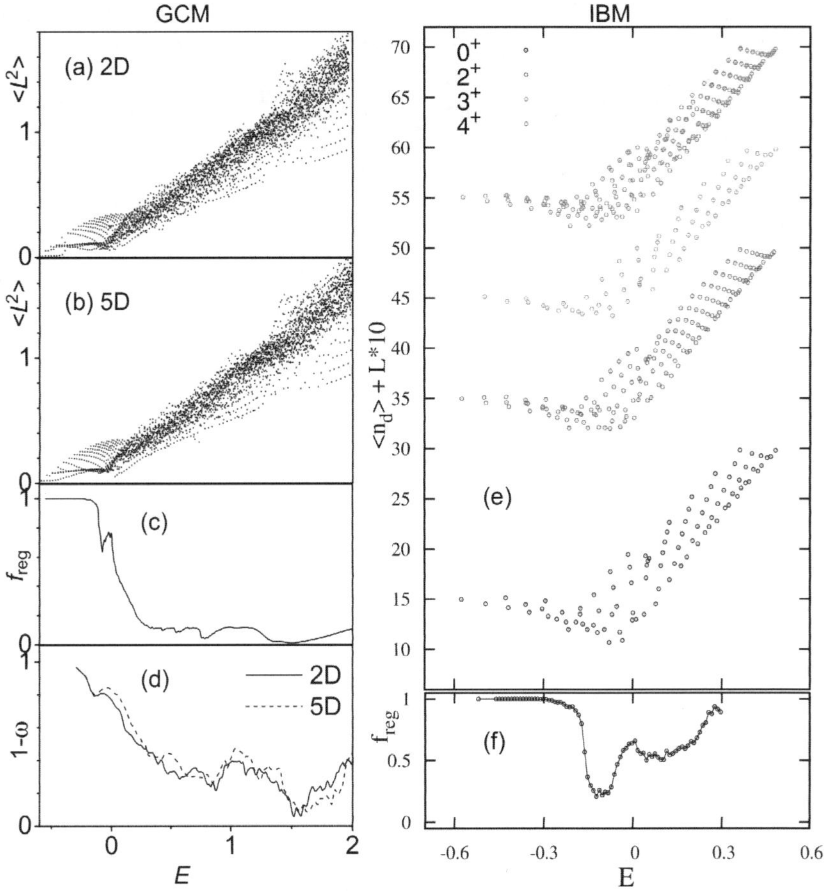

Figure 4. First column: Classical and quantum chaos of the GCM for $B = 0.62$. Peres lattices for $\kappa = 1 \cdot 10^{-4}$ are compared for 2D (a) and 5D (b) quantization. (c) Classical regular fraction of the phase space $f_{\rm reg}$ and (d) corresponding quantum counterpart—Brody parameter ω calculated for $\kappa = 2.5 \cdot 10^{-5}$. Second column: IBM dynamics at $\chi = -0.9$ and $\eta = 0.5$. (e) Peres lattices of the n_d operator for different angular momenta ($N = 30$ bosons), and (f) corresponding fraction of regularity $f_{\rm reg}$.

correspondence is apparent, though the quantum measures tend to overemphasize the regularity.

Figure 4 confronts the Peres lattices and the measures of chaos for both 2D and 5D quantization schemes. The lattices are displayed in panels (a) and (b), showing small differences in structure, but the distribution of regular and chaotic areas is equal. This is in agreement with the negligible difference between the 2D and 5D cases of $\omega(E)$ (panel d), which remains within the limits of the standard error in our calculations. Finally, observe the influence of the classicality parameter κ on the level density by comparing panel (a) with Figure 2 (d).

3. Interacting Boson Model

To demonstrate a different application of the Peres method we consider the Hamiltonian [1]

$$H = a\left[\frac{\eta}{N}\,n_d - \frac{1-\eta}{N^2}\,Q_\chi \cdot Q_\chi\right] \tag{4}$$

with n_d the d-boson number operator (N is a conserved total number of bosons), Q_χ the quadrupole operator, and (η, χ) two external parameters defining the Casten triangle. Its vertices $(\eta, \chi) = (0,0)$, $(1,0)$, and $(0, -\sqrt{7}/2)$ correspond to the O(6), U(5), and SU(3) dynamical symmetries, respectively. The system is integrable in the triangle vertices and along the side with $\chi = 0$. The Peres operator we can take any of the Casimir invariants connected with the symmetry groups, but our results are best demonstrated by the choice the U(5) operator n_d.

Recently, some low-energy rotational bands populating the spectrum below the saddle point energy in the axially-deformed part of the Casten triangle were explained analytically using specific SU(3)-like boson condensates [12]. In this contribution we turn to the highly excited states. Figure 4(e) shows the eigenstates of IBM in the arc of regularity in the interior of the Casten triangle, distributed into 4 Peres lattices according to their angular momentum. Apart from a few ordered low-lying states (present also in other parts of the parametric triangle), we may observe two distinct regular regions: one at $E \approx 0$ and the second reaching the highest accessible energy. Surprisingly, the lattices for different angular momentum L form the same pattern there. The analyses based on wave function correlation, $E2$ transitions and moments of inertia reveal that these "regular" energies associate into bands, showing signatures of SU(3) QDS in the $L(L+1)$ excitation energy pattern [13]. This behavior is strongly accentuated along the regular arc. While the low-energy dynamics are almost completely regular in the entire Casten triangle, the partially increased regularity of the arc (relative to the neighboring regions) is brought about essentially by the regular high-lying modes, which has already been shown [1] [this is also supported by the classical fraction of regularity $f_{\rm reg}$ for states with $L = 0$, drawn in the panel (f)]. It suggests a connection between quasi-dynamical symmetry and regularity.

4. Concluding remarks

We have calculated some measures of chaos for the classical and quantum versions of the geometric collective model (we dispose of two different quantum GCM with the same classical limit) and the interacting boson models in the wide range of external parameters and energy, observing good correspondence between the classical fraction of regularity and the Brody parameter that quantifies chaos in the quantum case. This correspondence is supported by the visual observation of the regular and chaotic structures in the Peres lattices.

In addition, we adopted the geometric method by Horwitz *et al* that estimates the position of regular-chaotic transition and follows the convex-concave shape change of the kinematically accessible area border in the GCM, which has been proven analytically [14]. We also adopted the method of $1/f^\alpha$ noise by Relaño *et al*, obtaining expected $\alpha = 2$ exponent in an integrable configuration and a correspondence in the energetic dependence of the measures $1-\omega$ and $\alpha - 1$ in a configuration with mixed dynamics.

In IBM we have observed a strong correlation between the variation of classical and quantum degrees of regularity and the occurrence rate of the rotational bands. This suggests that the separation of rotational motion from intrinsic vibrational dynamics is enhanced by the intrinsic regularity, while on the other hand, the chaotic intrinsic states are more prone to mixing, if the rotation comes into play; the regular states seem to be protected by their symmetry, in contrast. Different methods of rotational band identification are described in [13], being in full agreement with the results obtained by Peres method.

Acknowledgments

This work was supported by the Czech Science Foundation (202/09/0084), by the Czech Ministry of Education (MSM 0021620859), PAPIIT-UNAM (grant IN113808) and Conacyt, Mexico.

References

[1] Alhassid Y and Whelan N 1991 *Phys. Rev. Lett.* **67** 816
Whelan N and Alhassid Y 1993 *Nucl. Phys.* A **556** 42
[2] Cejnar P and Stránský P 2004 *Phys. Rev. Lett.* **93** 102502
[3] Stránský P, Kurian M and Cejnar P 2006 *Phys. Rev.* C **74** 014306
[4] Cejnar P and Stránský P 2009 *Phys. Rev.* E **78** 031130
[5] Rowe D J 2004 *Nucl. Phys.* A **745** 47
Rosensteel G and Rowe D J 2005 *Nucl. Phys.* A **759** 92
[6] Horwitz L, Zion Y B, Lewkowicz M, Schiffer M and Levitan J 2007 *Phys. Rev. Lett.* **98** 234301
[7] Gutzwiller M C 1990 *Chaos in Classical and Quantum Mechanics* (New York: Springer-Verlag)
[8] Peres A 1984 *Phys. Rev. Lett.* **53** 1711
[9] Relaño A, Gómez J M G, Molina R A, Retamosa J and Faleiro J 2002 *Phys. Rev. Lett.* **89** 244102
[10] Stránský P, Hruška P and Cejnar P 2009 *Phys. Rev.* E **79** 046202
Stránský P, Hruška P and Cejnar P 2009 *Phys. Rev.* E **79** 066201
[11] Macek M, Stránský P, Cejnar P, Heinze S, Jolie J and Dobeš J 2007 *Phys. Rev.* C **75** 064318
[12] Macek M, Dobeš J and Cejnar P 2009 *Phys. Rev.* C **80** 014319
[13] Macek M, Dobeš J and Cejnar P *Phys. Rev.* C in press
[14] Stránský P *et al* work in progress

Ultra-High-Energy Cosmic Ray and Neutrino Physics using the Moon

Olaf Scholten
KVI, University of Groningen, Groningen, The Netherlands
E-mail: Scholten@kvi.nl

Abstract. The intriguing mystery of ultra-high energy (UHE) cosmic particles is the nature of their sources. In this presentation we indicate how these UHE particles, in order to uncover their sources, can be detected using radio observations of the Moon. When high-energy cosmic rays impinge on a dielectric, such as the lunar regolith, radio waves are produced through the Askaryan effect. At wavelengths comparable to the length of the shower produced by an UHE cosmic ray or neutrino, radio signals are an extremely efficient way to detect these particles. First results of observations obtained with the Westerbork Synthesis Radio Telescope will be reported, already showing improved flux limits for neutrinos with energies in excess of $3\,10^{22}$ eV. The status of the planning of similar observations with LOFAR will be presented.

1. Ultra-high-energy cosmic rays

The flux of cosmic rays [1] falls off by about 32 orders of magnitude with energy increasing by about 11 orders of magnitude from 1 GeV to 10^{20} eV. The intriguing aspect is that at energies above 10^{10} eV the spectrum falls off almost like a straight line when plotted on a double logarithmic scale, indicating that the flux is a power law as function of energy E, falling off like $E^{-2.7}$. From this observation, indicating that the spectrum is non-thermal, it can immediately concluded that there are sources producing cosmic rays, even at the highest energies of 10^{20} eV. The highest energies are really large, the kinetic energy of a single proton at 10^{20} eV exceeds the kinetic energy of a tennis ball in a very good service. This energy is of such a magnitude that, even on astronomical scale, it is not possible to find objects where such energies can be reached. The usual mechanisms for accelerating UHE cosmic-rays involves the Fermi acceleration process which can occur only in the presence of magnetic fields which obey the Hillas criterium [2]; they have to be sufficiently strong over large enough distances to bend the particles in semi-circular orbits. At the highest energies there are no known objects that have strong enough magnetic fields over sufficient distances to obey this criterium and one has to explore other possible sources, see Ref. [3] for a discussion. The challenge is thus to locate the astrophysical sites where these particles originate from. At the Pierre Auger Observatory it has been observed that the arrival directions of cosmic rays at the highest energies show a strong correlation with the distribution of mass in the universe [4] which is what one would expect. A difficulty to measure these particles is presented by the extremely low flux of well below a single particle per km^2 per century. The detection method discussed here is shown to be very efficient to detect these small fluxes.

Protons at energies exceeding $6\,10^{19}$ eV may create pions when interacting with the cosmic microwave-background radiation through the so-called GZK process [5]. This suppresses the flux of UHE protons, evidence of which is seen in recent data [6], while creating neutrinos. It is also very likely that neutrinos are generated at the production sites of the energetic cosmic rays. This, compounded with the fact that neutrinos are moving on straight lines (unlike charged cosmic rays which are deflected by intergalactic magnetic fields) may make neutrinos much more effective as a signal directing towards the astrophysical sources, which could be extremely powerful cosmic accelerators or, more exotically, decaying super-massive dark-matter particles or topological defects. Recent limits on the photon flux at high energies [7] have made the latter scenario less likely. Since neutrinos interact only weakly and since their flux at the highest energies is extremely low we need exceedingly large and efficient detector systems to measure them. In the following section it will be shown that the Moon, with an area of $2\,10^7\,\text{km}^2$, can be used as a highly efficient detector [8] for ultra-high energy neutrinos.

2. Principle of particle detection using the Moon

In the interaction of an UHE neutrino with the material in the lunar regolith, the top layer of the Moon consisting of dust and small rocks, about 20% of the energy of the neutrino is converted into a cascade of particles, called the hadronic shower, while the remaining energy is carried by a single lepton which will not significantly contribute to the process considered here. In the hadronic shower there are large numbers of energetic hadrons, leptons, and photons all moving through the regolith with the light velocity c. Due to processes like Compton scattering and positron collisions off electrons bound in the atoms of the regolith a net excess of electrons is formed in the shower. This excess is typically of the order of 20% of the number of particles in the shower, $N \gg 10^6$. This large charge cloud is moving with almost the light velocity through a medium where the propagation velocity of electromagnetic waves is considerably less (index of refraction $n = 1.8$) and therefore Cherenkov radiation is emitted.

At wavelengths that are larger than the typical dimension of the charge cloud (10 cm) the emitted Cherenkov radiation is coherent, a process known as the Askaryan effect [9]. The intensity of this radiation reaches a maximum at a frequency of about 3 GHz where the wave length is of the same order of magnitude as size of the charge cloud. At this frequency the emitted radiation is strongly concentrated at the Cherenkov angle since the distance over which the shower extends, about 3 m, is much larger than the wave length. At lower frequencies the angular spread of the emitted radiation increases. At frequencies of 100-200 MHz the wavelength is comparable to the longitudinal extent of the shower which gives rise to a large angular spread of the emitted radiation. The increase of the angular spread is important for two reasons, one is that for many more directions of the incoming cosmic ray the signal will be detectable at Earth. A second, even more important reason is that for a shower parallel to the lunar surface the Cherenkov emission angle is the same as the angle of total internal reflection at the surface. As a result only the cosmic rays impinging at the rim of the moon, directed towards the Earth, emit a detectable signal at high frequencies while at the lower frequencies the total lunar surface may emit detectable signals [8]. A major advantage for the detection of neutrinos on the Moon is the rather long attenuation length of $\lambda_r = (9/\nu[\text{GHz}])$ m for radio waves which makes a very large detection volume. As a result the detection efficiency increases with about the third power of the wave length.

The idea of measuring cosmic rays through their impacts on the Moon was first proposed in Ref. [10] while the first experiments were carried out with the Parkes telescope [11], later followed by others [12, 13]. These observations are all performed at relatively high frequencies (2 GHz) where the emission is strongest. It should be noted that the radio-emission process for cosmic rays in the Earth atmosphere is governed by another mechanism. The magnetic field of the Earth induced electric current in the shower plasma and radio waves are emitted due to the changing magnitude of this electric current [14] while the Askaryan effect plays only a secondary role. Similar to the case of radiation from the Moon, the frequency spectrum can be understood by considering the important length scales in the problem [15].

3. WSRT Observations

The emission of 3 m radio waves from impacts of high energy neutrinos on the Moon is exploited in our observations with the Westerbork Synthesis Radio Telescope (WSRT) and has resulted in the most stringent flux limit at the highest energies [16]. The WSRT consists of an array of 14 parabolic antennas of 25 m diameter. Only 11 of the 12 equally spaced WSRT dishes are used for this experiment. In the observations we have employed the Low Frequency Front Ends (LFFEs) which cover the frequency range 115–180 MHz with full polarization sensitivity, sampled as 8 subbands of 20 MHz each by the Pulsar Machine II (PuMa II) backend [17]. We have used 4 frequency bands centered at frequencies of 123, 137, 151 and 165 MHz for each of two different beams aimed at different sides of the Moon. This creates the possibility of an anti-coincidence trigger since a lunar Cherenkov pulse should only be visible in only one of the two beams. The total bandwidth per beam is 65 MHz. The time series data is recorded for each subband with a sampling frequency of 40 MHz. The data is processed in blocks of 0.1 s, where each block is divided in 200 traces of 20,000 time samples.

In a first processing stage the data is searched for large pulses after applying the following two corrections. First the narrow band Radio Frequency Interference (RFI) is filtered from the data. All frequency channels exceeding a smooth background by 50% or more are marked as RFI lines and set equal to zero. Second, the dispersion due to the ionosphere of the Earth is eliminated. The vertical total electron content (TEC) values of the ionosphere is corrected for the elevation of the Moon to obtain the slanted-TEC (STEC) value. Because of variations of the ionosphere on short timescales, in the analysis we account for an inaccuracy in the determined STEC value which will result in an increased time width of the pulses.

From the de-dispersed and RFI-mitigated spectrum a 5-time-sample-summed relative-power spectrum is constructed,

$$P_5 = \frac{\sum\limits_{5\ \text{samples}} P_x}{\left\langle \sum\limits_{5\ \text{samples}} P_x \right\rangle} + \frac{\sum\limits_{5\ \text{samples}} P_y}{\left\langle \sum\limits_{5\ \text{samples}} P_y \right\rangle}, \tag{1}$$

where the averaging is done over one time trace (20,000 time samples), and x and y denote the two polarizations. The summation over 5 time samples corrects for a possible error in the STEC value. A trigger is generated when in all four frequency bands in the same beam a value $P_5 > 5$ is found. The first and last 250 time samples are excluded since the RFI noise is not properly eliminated from these parts of the spectrum. The value S is defined as the sum over the maximum P_5 values in the 4 frequency bands, $S = \sum_{4\ \text{bands}} P_5$.

In the subsequent off-line analysis the data is 'cleaned' by eliminating a timer signal that repeats with a frequency of 102.4 s^{-1}, pulses that are very wide, and pulses that appear in both beams in the same time trace. The largest remaining pulse in the data has a value of $S = 76$. On the basis of simulations we can show that this observation translates into being able to exclude pulses coming from the moon with a strength exceeding $S > 120$ (corresponding to $120\sigma^2 \times 5 = 240$ kJy) over a period of 40 h observation [16]. The lack of pulses of a certain power converts into a limit on the flux of neutrinos where attenuation of the signal in the Moon, transmission at the lunar surface, and angle with respect to the direction of the neutrino have to be taken into account [8]. The resulting 90% confidence limit flux limit shown in Fig. 1 assuming that the neutrino cross sections given in Ref. [18]. The limit is still well above the Waxman-Bahcall limit [19] but borders on the predictions of a top-down model [20] for exotic

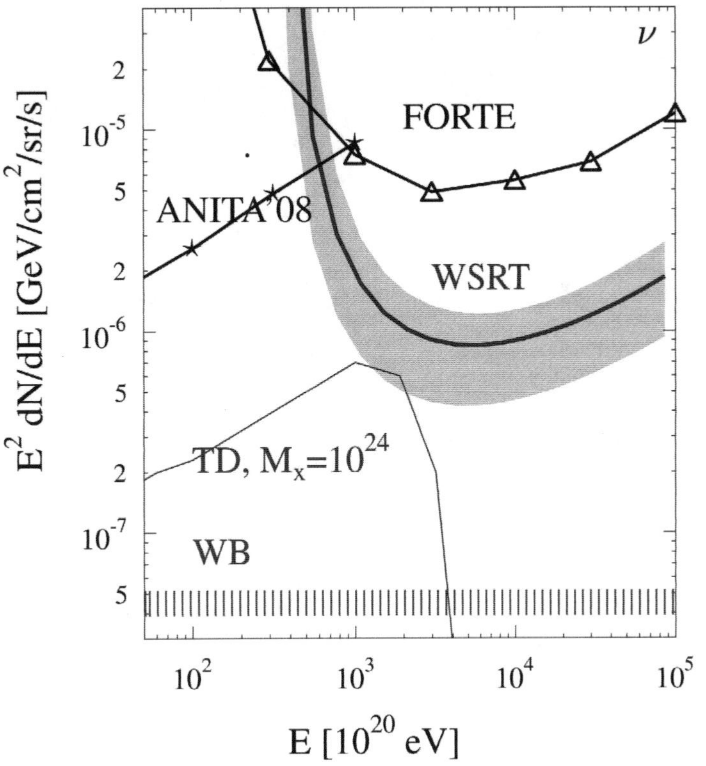

Figure 1. The limit [16] on the flux of UHE neutrinos obtained from observations with the WSRT is compared with those of other experiments and some model calculations, see text.

particles of mass $M_X = 10^{24}$ eV. The previous limits in the UHE region have been set by ANITA [21] and FORTE [22].

The determined limits are subject to systematic errors on the acceptance simulation which are due to uncertainties in the density (10% in threshold energy), the attenuation length (40% in flux), and the stopping power of the regolith, (20% in flux). The error in the moon coverage of the two beams is estimated at 20%. Adding these errors in quadrature gives a systematic error on the flux of 50% as indicated in Fig. 1.

4. Future

LOFAR (Low Frequency Array) is a new concept radio telescope that is presently being rolled-out in The Netherlands and other European countries and is also used for the detection of cosmic rays [23, 24]. LOFAR couples a large number (thousands) of simple wire dipole antennas. Via software control the relative phasing of the antennas can be tuned such that the array is (virtually) pointed towards a certain direction in the sky. The fact that this pointing is done in software without any physical movement of antennas makes the system very flexible and gives the possibility to simultaneously perform observations with multiple beams.

Part of our ongoing activity in this field concerns the optimization of the neutrino observations with LOFAR [25]. For the observation of neutrino impacts on the moon the multi-beam option, combined with the large collecting area, is very important. The large area improves the signal-to-

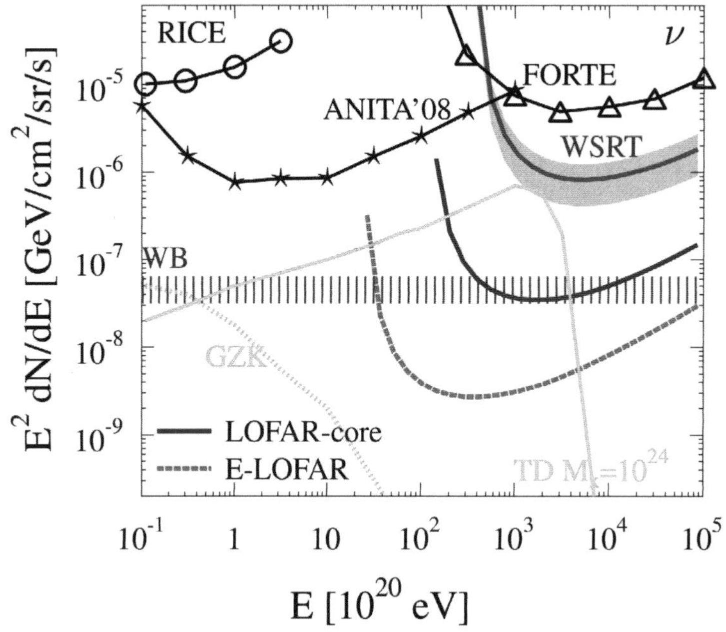

Figure 2. The predicted sensitivities for two different LOFAR configurations (see text) are compared with the results obtained from the NuMoon observations with WSRT [16].

noise ratio thus increasing the detection sensitivity at lower energies while the multi-beam option allows for the formation of sufficiently many beams to cover the complete lunar surface. The position resolution on the moon, offered by the large baselines, further improves the sensitivity to genuine pulses generated by neutrino impacts. In Fig. 2 the sensitivity is indicated that can be reached with a week of observation using only the stations in the core (blue curve) while the sensitivity attained by using all LOFAR stations, including the international ones, is such that in one month accumulated time should result in a large number of observed events for a neutrino flux at the level of the Waxman-Bahcall limit.

The large detection efficiency of LOFAR opens the possibility to observe neutrino absorbtion lines. UHE neutrinos will be absorbed on the low-energy cosmic neutrino background (CνB) through the excitation of the Z boson resonance. This annihilation process is expected to lead to sizable absorption dips in the UHE neutrino spectra [26]. The positions of these absorption lines depend on the red-shift of the source emitting the UHE neutrinos, z_s. An exciting alternative is that the neutrino mass is not constant but varies with red-shift, mass varying neutrinos (MaVaN) [27], proposed as a model for dark energy. Such a neutrino mass variation introduces a structure of the absorption lines which is distinctly different to that obtained for red-shift independent ν-masses.

The magnitude of a radio pulse created by the impact of neutrinos of a given energy on the Moon depends on the viewing angle and the distance below the lunar surface of the shower. The pulse-height is thus not a direct measure of the neutrino energy but the measured pulse-height spectrum does reflect the energy distribution of the neutrinos as has been investigated in Ref. [28]. The predicted pulse-height spectra for different scenarios for the neutrino mass and various red-shifts for the neutrino sources is shown in Fig. 3, indicating that with a sensitive detection system one should be able to distinguish sources at z_s=50 from more nearby ones.

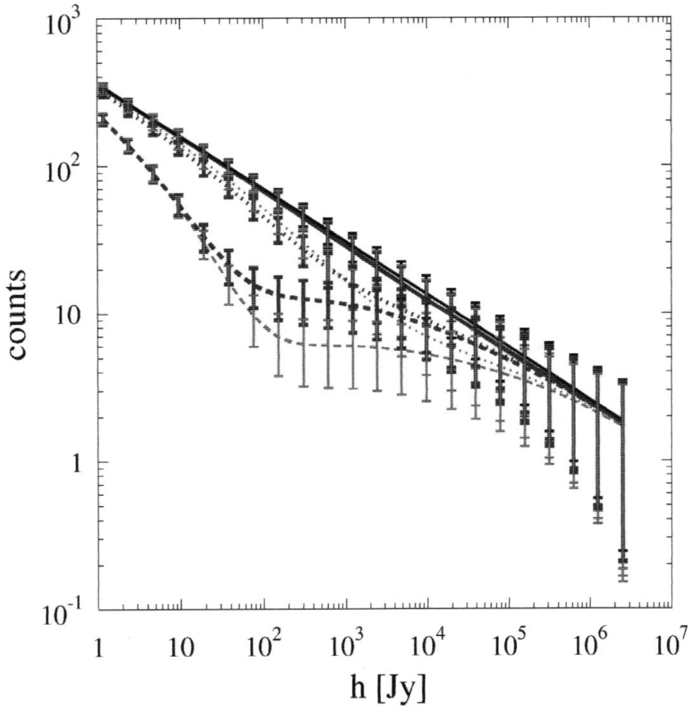

Figure 3. The expected count rate [28] for a realistic experiment for the different neutrino absorption dips for sources at various red-shifts z_s=5 (drawn line), z_s=20 (dotted line), and z_s=50 (dashed line). The thick, blue, curves give the absorption spectra for the case where the neutrino mass is independent of red-shift, while the thin, red, curves gives the predictions in the MaVaN model.

Acknowledgments
This work was performed as part of the research programs of the Stichting voor Fundamenteel Onderzoek der Materie (FOM), with financial support from the Nederlandse Organisatie voor Wetenschappelijk Onderzoek (NWO). I wish to thank all members of the NuMoon collaboration. Without their contributions it would not have been able to perform the observations summarized in this work.

References

[1] Amsler C *et al* 2008 *Phys. Lett.* B **667** 1
Gaisser T K 2006 *J. Phys.: Conf. Series* **47** 15
[2] Hillas A M 1984 *Ann. Rev. Astron. Astrophys.* **22** 425
[3] Torres D F and Anchordoqui L A 2004 *Rep. Prog. Phys.* **67** 1663 (*Preprint* astro-ph/0402371v3)

[4] Abraham J *et al* 2007 [Pierre Auger Collaboration] *Science* **318** 938
Abraham J *et al* 2008 [Pierre Auger Collaboration] *Astropart. Phys.* **29** 188
[5] Greisen K 1966 *Phys. Rev. Lett.* **16** 748
Zatsepin G and Kuzmin V 1966 *Pis'ma Zh. Eksp. Teor. Fiz.* **4** 114
[6] Abraham J *et al* 2008 [Pierre Auger Collaboration] *Phys. Rev. Lett.* **101** 061101
[7] Abraham J *et al* 2008 [Pierre Auger Collaboration] *Astropart. Phys.* **29** 243
Abraham J *et al* 2009 [Pierre Auger Collaboration] *Astropart. Phys.* **31** 399
[8] Scholten O *et al* 2006 *Astropart. Phys.* **26** 219
[9] Askaryan G 1962 *Sov. Phys. J.E.T.P.* **14** 441
[10] Dagesamanskii R and Zheleznyk I 1989 *Sov. Phys. J.E.T.P.* **50** 233
[11] Hankins T, Ekers R and O'Sullivan J 1996 *Mon. Not. Roy. Astron. Soc.* **283** 1027
[12] Gorham P *et al* 2004 *Phys. Rev. Lett.* **93** 41101
[13] Beresnyak A *et al* 2005 *Astron. Rep.* **49** 127
[14] Scholten O, Werner K and Rusydi F 2008 *Astropart. Phys.* **29** 94
Werner K and Scholten O 2008 *Astropart. Phys.* **29** 393
[15] Scholten O and Werner K 2009 *Nucl. Instrum. Meth.* A **604** S24
[16] Scholten O *et al* 2009 *Phys. Rev. Lett.* **103** 191301
Buitink S *et al A&A* in press
[17] Karuppusamy R, Stappers B and Van Straten W 2008 *Publ. Astron. Soc. Pacific* **120** 191 (*Preprint* arXiv:0802.2245)
[18] Gandhi R 2000 *Nucl. Phys.* B **91** 453
[19] Waxman E and Bahcall J N 1999 *Phys. Rev.* D **59**, 023002
[20] Protheroe R and Stanev T 1996 *Phys. Rev. Lett.* **77** 3708
[21] Gorham P *et al* 2009 *Phys. Rev. Lett.* **103**, 051103
[22] Lethinen H *et al* 2004 *Phys. Rev.* D **69** 013008
[23] Hörandel J R *et al* 2009 *Nucl. Phys.* B **196** 289
[24] Horneffer A *et al* 2009 *Nucl. Instrum. Meth.* A **604** S20
Horneffer A *et al* 2010 *Nucl. Instrum. Meth.* A **617** 482
[25] Singh K *et al* 2009 *Proc.* 31st *ICRC*
[26] Ringwald A and Schrempp L 2006 *J. Cosmol. Astropart. Phys.* **0610** 012
[27] Fardon R, Nelson A E and Weiner N 2004 *J. Cosmol. Astropart. Phys.* **0410** 005
[28] Scholten O and Van Vliet A 2008 *J. Cosmol. Astropart. Phys.* **0806** 015

Journal of Physics: Conference Series **239** (2010) 012004 doi:10.1088/1742-6596/239/1/012004

Nuclear reaction rates and opacity in massive star evolution calculations

D Bahena[1], J Klapp[2,3] and H Dehnen[3]

[1] Astronomical Institute of the Academy of Sciences, Boční II 1401, 14131 Praha 4, Czech Republic

[2] Instituto Nacional de Investigaciones Nucleares, Km. 36.5 Carr. México-Toluca, 52750 Edo. de México, México

[3] Universität Konstanz, Fachbereich Physik, Fach M568, D-78457 Konstanz, Germany

E-mail: `jaime.klapp@inin.gob.mx`

Abstract. Nuclear reaction rates and opacity are important parameters in stellar evolution. The input physics in a stellar evolution code determines the main theoretical characteristics of the stellar structure, evolution and nucleosynthesis of a star. For different input physics, in this work we calculate stellar evolution models of very massive first stars during the hydrogen and helium burning phases. We have considered 100 and $200 M_{\odot}$ galactic and pregalactic stars with metallicity $Z = 10^{-6}$ and 10^{9}, respectively. The results show important differences from old to new formulations for the opacity and nuclear reaction rates, in particular the evolutionary tracks are significantly affected, that indicates the importance of using up to date and reliable input physics. The triple alpha reaction activates sooner for pregalactic than for galactic stars.

1. Introduction

The first stars provided the initial heavy elements enrichment of the intergalactic medium (IGM). Massive primordial stars marked the first step in the chemical evolution of the universe. The abundance of the CNO elements are fundamental, they form the bulk of the heavy elements with important roles in stellar interior opacities, and energy generation, affecting the star's lifetime, evolutionary track on the HD diagram, and heavy element yields.

Two major factors that affect the nucleosynthetic yields are the $^{12}C(\alpha, \gamma)^{16}O$ reaction rate and the efficiency of semiconvective mixing (Weaver and Woosley 1993). The evaluation of the cross-section for the different nuclear burning processes is required for establishing the energy balance of the star. Many cross-sections are known with high accuracy. However, there are uncertainties with some important processes for both hydrogen, and helium burning, where nuclear rates are not well known (Cassisi 2005). The radiative opacity is related to the mean free path of the photons and it is very important in determining the efficiency of heat transfer via radiative processes.

Nuclear reaction rates and opacities have been changing with time. In this paper, we calculate models of very massive first stars with the aim of comparing their evolutionary tracks by using different old and new formulations for the opacity and nuclear reaction rates.

In Section 2 we present a brief review concerning nuclear reaction rates. In Section 3, the same for the opacity as well. In Section 4 some results are described, and in Section 5 we discuss the importance of the input physics, specifically nuclear physics, and outline some conclusions.

Journal of Physics: Conference Series **239** (2010) 012004 doi:10.1088/1742-6596/239/1/012004

2. Nuclear reaction rates

2.1. Nuclear reactions in astrophysics

Nuclear processes in massive stars have been proposed as the source of most of the isotopes heavier than helium now present in the universe. Stellar evolution models incorporate detailed nuclear reaction networks based on accurate calculations or measurements of nuclear cross-sections.

A first review of the available experimental data on cross-sections for the nuclear interactions of neutrons, protons, and alpha particles with a number of light and intermediate-mass nuclei was made by Fowler, Caughlan and Zimmermann (1967).

Stellar thermonuclear reaction rates were revised and updated by Fowler, Caughlan and Zimmermann (1975), and Harris *et al* (1983) adding new laboratory measurements of nuclear cross-sections. Furthermore, new data were included by Caughlan *et al* (1985), and Caughlan and Fowler (1988).

Remarkable progress have been made in reaction rates at energies close to those of astrophysical relevance (Rolfs and Rodney 1988). Despite these efforts, important uncertainties remain.

The Nuclear Astrophysics Compilation of Reaction Rates (NACRE) have documented and evaluated sets of experimental data or theoretical predictions for a large number of astrophysical nuclear reactions (Angulo, Arnould and Rayer 1999). This compilation included the rates for all the charged-particle-induced reactions involved in the pp, CNO, NeNa and MgAl chains. The first two chains are energy producers but all four are important nucleosynthesis agents. Non-explosive He-burning reactions were also included.

2.2. Evolutionary calculations

The abundances of the isotopes from C to Al produced by the non-explosive CNO, NeNa and MgAl modes of hydrogen burning, as well as by helium burning were calculated by Arnould, Goriely and Jorissen (1999) using the NACRE thermonuclear rates.

In their evolutionary calculations of zero-metallicity stars, Marigo *et al* (2001) evaluated the changes in chemical abundances due to nuclear reactions by solving a network consisting of a set of reactions for the pp1 chain, the CNO tri-cycle and the most important α-capture reactions. The reaction rates are from Caughlan and Fowler (1988) compilation, except for $^{17}\mathrm{O}(p,\alpha)^{14}\mathrm{N}$ and $^{17}\mathrm{O}(p,\gamma)^{18}\mathrm{F}$, for which they used determinations by Landré *et al* (1990).

To study the cosmological effects of the first stars, Tumlinson, Shull and Venkatesan (2003) considered a nuclear reaction network coupled to the stellar structure equations. The network includes the relevant nuclear reactions for 24 isotopic abundances. The reaction rates are from Caughlan and Fowler (1988) compilation.

A nuclear network following the evolution of 13 isotopes was implemented by Claret (2004) for the pp-chains, CNO tri-cycle, helium and carbon burning reactions. The basic nuclear reactions are from Caughlan and Fowler (1988), while the screening factors are taken from Graboske *et al* (1973). In the new grids of stellar models from Claret (2005) all reaction rates are taken from NACRE with the exception of the $^{14}\mathrm{N}(p,\gamma)^{15}\mathrm{O}$ reaction, for which a new rate has been implemented (Runkle 2003, Formicola et al. 2004). This reaction is the slowest step of the CNO cycle and the effect on the evolutionary life of the stars will depend on the stellar mass.

The reliability of theoretical predictions of the evolutionary lifetime depends on the accuracy of the nuclear reaction rates since nuclear burning provides the bulk of the stellar luminosity during the main evolutionary phases (Cassisi 2005).

For calculating their first stars structure and evolution models, Bahena and Klapp (2010) used the NACRE compilation of nuclear reaction rates.

3. Radiative opacities

3.1. Tabulated opacities

The calculation of realistic stellar opacities is among the most difficult problem in stellar astrophysics. At the present time, the most commonly used opacities for stellar mixtures are those generated at the Los Alamos National Laboratory (LANL) and at the Lawrence Livermore National Laboratory. Opacities are available in tabular form and include many stellar mixtures with opacities computed over a wide density and temperature range.

In practice, extensive tables or, sometimes, an analytic fit to the tables are used that gives radiative and conductive opacities over a wide temperature and density range for the various compositions of interest. A specific opacity is obtained by a multidimensional interpolation between the tables. For a decades, the Cox and Stewart (1965, 1970a, 1970b) opacity tables were used as standard. To recompute new opacity tables, two major independent efforts were made. One was the Opacity Project (Seaton 1987, Berrington *et al* 1987), and the other one is known as OPAL.

The Rosseland mean opacity tables calculated with the OPAL code for various compositions over a wide range of densities and temperatures for typical stellar conditions were presented by Rogers and Iglesias (1992). The OPAL code removes several approximations in the equation of state and atomic data generation present in past calculations. A comparison to Eggleton, Faulkner and Flannery (1973) produces small differences. OPAL results allow accurate interpolation in temperature, density, hydrogen mass fraction, as well as metal mass fraction. The tables do not include atmospheric and high-temperature opacities, only radiation processes are taken into account.

Iglesias and Rogers (1991), and Rogers and Iglesias (1992) demonstrated that the inclusion of more metal lines that were previously considered can greatly increase the total opacity at temperatures above 10^5 K. Further refinements of the atomic physics were incorporated into the OPAL opacity code. The tables were extended to lower and higher densities as well as higher metallicity (Iglesias, Rogers and Wilson 1992).

3.2. Opacity for massive stars

One of the major problems in working with very massive stars is that a quasi-static stellar evolution code very easily becomes unstable. The problem becomes worse as the mass of the star is increased and in many cases it is very difficult to find out what is causing the instabilities. The use of opacity tables contribute to make the code unstable.

During many years the opacity was obtained by interpolating from Cox-Stewart opacity tables. The way in which the interpolation is done causes the opacity to oscillate slightly as you move from the vicinity of one table point to another. It is this small variation in the opacity which makes the code unstable, which also shows how sensitive very massive star calculations are to the programming input physics. In view of this difficulty it is convenient to use Christy's (1966) analytic fit to the tables which can be written as

$$\begin{aligned}
\kappa &= P_e\Big[5.4\times10^{-13}\rho^{-1}T_4^{-1}(1/(1+2T_9)) \\
&+ XT_4^{1/2}(2\times10^6T_4^{-4}+2.1T_4^{-6})^{-1} \\
&+ X\{4.5T_4^6+T_4^{-1}(4\times10^{-3}T_4^{-1}+2\times10^{-4}\rho^{-1/4})^{-1}\}^{-1} \qquad (1)\\
&+ Y\{(1.4\times10^3T_4+T_4^6)^{-1}+1.5(10^6+0.1T_4^6)^{-1}\} \\
&+ (1-X-Y)\{T_4^{1/2}+(20T_4+5T_4^4+T_4^5)^{-1}\}\Big],
\end{aligned}$$

where T is the temperature, $P_{\rm e}$ the electron pressure, and the first term has been modified to take into account the Klein-Nishina contribution to the opacity (Klapp 1982, 1983). The first term gives the free-electron scattering contribution, the second is due to hydrogen, the third to helium, and the last one to heavy elements.

For very massive stars most of the instabilities disappear with the use of equation (1), which has been used by Klapp (1982, 1983, 1984), Klapp *et al* (2005, 2006), Bahena (2006), and Bahena, Klapp and Dehnen (2007).

3.3. Evolutionary models

The computation of new opacity tables was performed by the Geneva group with the Los Alamos opacity programme (Maeder 1990).

The new stellar opacities are found to lead to revised evolutionary models for stars from intermediate to high masses. With the use of these opacities a remarkably close agreement with observed stars was obtained (Stothers and Chin 1991). Claret and Giménez (1991) used opacities based in the latest calculations from Los Alamos. The chemical composition adopted corresponded to Population I. The new opacities obtained at Lawrence Livermore National Laboratory by Rogers and Iglesias (1992) were introduced by Claret and Giménez (1992) together with the Los Alamos Astrophysical Opacity Library (LAOL) (Huebner *et al* 1977, Weiss, Keady and Magee 1990).

Schaller *et al* (1992) presented new grids of stellar models at various metallicities using the tables of Huebner *et al* (1977), the opacities by Rogers and Iglesias (1992) and by Kurucz (1991) at low temperature. Charbonnel *et al* (1993), and Schaerer (1993a, 1993b) also used the OPAL radiative opacities (Iglesias, Rogers and Wilson 1992). Claret (1995) adopted the OPAL opacities and Los Alamos tables. Claret (2004, 2005) adopted in his models the radiative opacities by Iglesias and Rogers (1996) for higher temperatures, complemented with Alexander and Ferguson (1994) results for lower temperatures.

The radiative opacities from the OPAL group (Rogers and Iglesias 1992, Iglesias and Rogers 1993) were used by Marigo *et al* (2001). The conductive opacities for electron-degenerate matter were from Hubbard and Lampe (1969). In the domain of high temperatures, Siess, Livio and Lattanzio (2002) used the OPAL opacity tables and other opacities computed by several authors.

For a 300 $M_\odot$ star, Bromm, Kudritzki and Loeb (2001) considered that the only source of opacity is electron scattering and used the well known expression $\kappa = 0.20(1+X)$ cm^2 g^{-1}. The Rosseland mean opacity from the OPAL tables (Iglesias and Rogers 1996) have been used by Tumlinson, Shull and Venkatesan (2003) to calculate their models for very massive first stars.

For the structure and evolution models of the first stars, Bahena and Klapp (2010) used the opacities from the OPAL group.

4. Results

4.1. Initial conditions

For this work we calculated models for 100 and 200$M_\odot$ galactic and pregalactic Population III stars, and two different chemical compositions which are the following: $(X, Z) = (0.765, 10^{-6})$, and $(X, Z) = (0.765, 10^{-9})$, for galactic and pregalactic Population III stars, respectively.

4.2. Code and input physics

The system of stellar structure and evolution equations used in the numerical code has been described by Klapp (1982, 1983) and Klapp et al. (2005, 2006). Concerning input physics, in this work we consider three cases: a) nuclear reaction rates from Harris *et al* (1983), and an analytic opacity formulation from Christy (1966); b) nuclear reaction rates from Harris *et al* (1983), and opacities from Rogers and Iglesias (1992) and Iglesias and Rogers (1993, 1996); c) the NACRE compilation from Angulo, Arnould and Rayer (1999) and opacities from the OPAL group. Evolutionary calculations have been made without mass loss and no rotation, during the hydrogen and helium burning phases.

Journal of Physics: Conference Series **239** (2010) 012004 doi:10.1088/1742-6596/239/1/012004

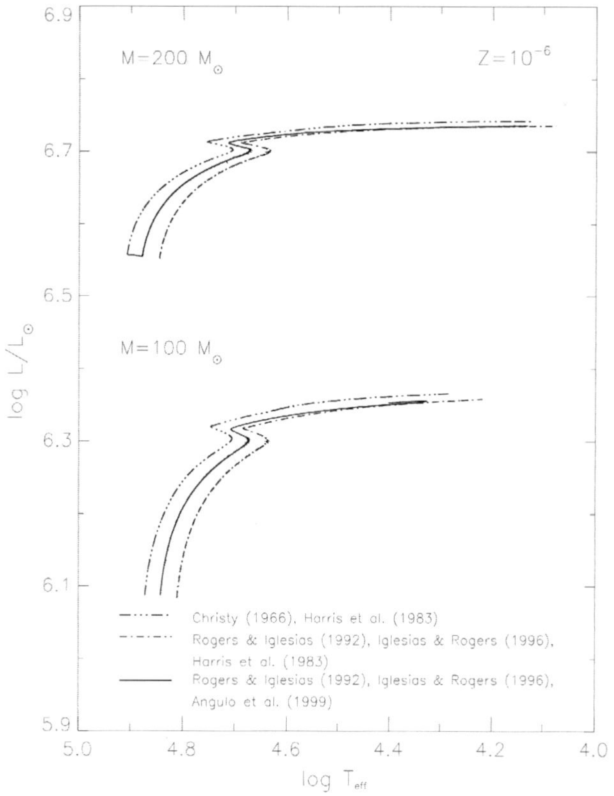

Figure 1. Evolutionary tracks in the HR-diagram for 100 and $200 M_\odot$ galactic stars with metallicity $Z = 10^{-6}$ during the hydrogen, and helium burning phases.

4.3. Evolutionary tracks

Figures 1 and 2 show the evolutionary tracks in the Hertzprung-Russell (HR) diagram for 100 and $200 M_\odot$ galactic and pregalactic Population III stars, respectively, with different input physics during the hydrogen and helium burning phases.

Massive star evolutionary tracks during the hydrogen and helium burning phases are characterized by a continuous movement to the right of the HR-diagram, describing different paths depending on the stellar mass and the mass loss rate.

In the HR-diagram the locus of very massive Population III stars is in the left upper part. Pregalactic stars are hotter than galactic ones. Stars with lower metallicity are shifted to the left because they are bluer than the others.

The most massive stars are very hot and luminous. Luminosity decreases with lower metallicity. Stars settle down on the main sequence with a high effective temperature and luminosity. During hydrogen burning all stars increase their luminosity while a helium mass fraction is formed. The central density and temperature both increases. At the end of hydrogen burning a transition to helium burning takes place.

For galactic stars the transition to helium burning is accompanied with an explosive event and the contraction of the star. For pregalactic stars, this transition occurs very smoothly because they are hotter and their central temperature is high enough to ignite helium promptly.

The hydrogen and helium burning phases takes place at the blue side of the HR-diagram. When stars evolve they move towards the red and without mass loss they do not experience the asymptotic giant branch (AGB).

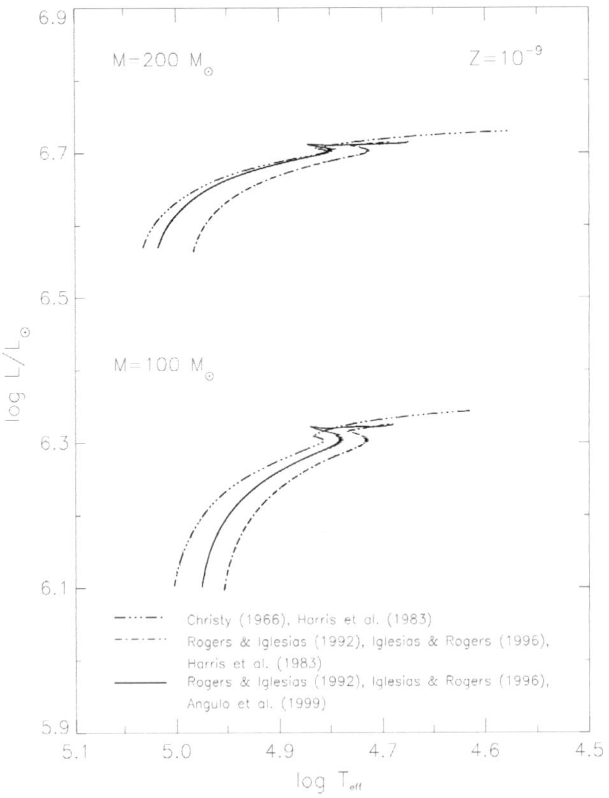

Figure 2. Evolutionary tracks in the HR-diagram for 100 and $200M_\odot$ pregalactic stars with metallicity $Z = 10^{-9}$ during the hydrogen, and helium burning phases.

Comparing the three studied cases for 100 and $200M_\odot$ galactic stars we observe the following: for case (a) (old opacity and old rates), stars are hotter and more luminous than for case (c) (new opacities and new rates) during the hydrogen and helium burning phases, and stars evolving without mass loss never reach the AGB. For case (b) (new opacities and old rates), stars evolve with lower effective temperature and luminosity than for case (a) during hydrogen burning, but similarly for helium burning.

For pregalacic stars of the same mass, the behavior is similar to galactic ones with the difference that stars are hotter and slightly less luminous than for the galactic case.

Theoretical analysis predict the final fate of massive stars. If the first generation of stars have been quite massive ($100 - 300M_\odot$), and retain their high mass until death, such stars will make pair-instability supernovae (PISNe) (Fryer, Woosley and Heger (2001). This is the main characteristic of these stars.

Heger and Woosley (2002) have explored helium cores nucleosynthesis in the mass range $M_{\rm He} = 64 - 130M_\odot$, corresponding to main sequence masses of approximately $140 - 260M_\odot$. They suggest different scenarios. The ultimate fate of a metal-free star depends critically on its mass. PISNe nucleosynthesis varies greatly with the helium core mass.

In our models, the stars develop large convective cores. The PISNe scenario and hypernovae (HNe) hypothesis have been discussed by Klapp and Bahena (2008).

The nucleosynthesis constrains on a pregalactic generation of stars has been studied by Abia *et al* (2001). They found that the possible contribution of the Population III remnants to the baryonic dark matter is insignificant.

It is often suggested that the initial heavy element enrichment is due to the first generation of stars. In our very massive models, the most abundant nuclei produced during hydrogen burning nucleosynthesis is ^{14}N (Bahena, Klapp and Dehnen 2007).

5. Discussion and conclusions

From the comparison of the evolution of very massive Population III stars using different opacities and nuclear reaction rates we found that the evolutionary tracks in the HR-diagram are similar. Pregalactic stars are hotter than galactic ones because the triple alpha reaction actives sooner during hydrogen burning.

However, there exist differences in the effective temperature and luminosity for the old and new values. The best results are obtained for the later case.

This shows that the input physics is very important in stellar evolution calculations. Different opacities and/or nuclear reaction rates yield different evolutionary tracks.

Uncertainties in several reaction rates still exist, such as the triple alpha reaction and the $^{12}C(\alpha,\gamma)^{16}O$ reaction. For this reaction NACRE considers a factor of 2 to be the uncertainty in the reaction rate, responsible for the ^{16}O abundance.

This reaction has a significant influence on massive star advanced stages of evolution. Heger *et al* (2002) have studied the nucleosynthesis and nuclear reaction uncertainties for massive star evolution. They consider that the uncertain $^{12}C(\alpha,\gamma)^{16}O$ reaction rate has a significant influence on advanced massive star evolution, and also determines whether a neutron star or black hole is formed.

For the critically important $^{12}C(\alpha,\gamma)^{16}O$ reaction, the cross-section used by Weaver and Woosley (1993), Woosley and Weaver (1995), and Woosley, Langer and Weaver (1993) has been 1.7 times the value given at each temperature by Caughlan and Fowler (1988). In their pre-supernova evolution and supernova explosion calculations, Umeda, Nomoto and Nakamura (2000) have chosen a value which is 1.4 times that given by Caughlan and Fowler (1988). Heger *et al* (2002) used a value which is 1.2 times the $^{12}C(\alpha,\gamma)^{16}O$ rate of Buchmann (1996).

In Marigo *et al* (2001) the uncertain $^{12}C(\alpha,\gamma)^{16}O$ rate was set to 1.7 times the value given by Caughlan and Fowler (1988) and Weaver and Woosley (1993) for a massive star nucleosynthesis calculation, with electron screening factors for all reactions taken from Graboske *et al* (1973).

The $^{12}C(\alpha,\gamma)^{16}O$ and $^{4}He(2\alpha,\gamma)^{12}C$ reactions compete for helium consumption, and thus the relative rate of these reactions determine whether oxygen or carbon is the dominant helium burning product. The triple alpha reaction rate is crucial for the first stars evolution but its rate is still under discussion.

Acknowledgments

We would like to thank KanBalam-UNAM and the Instituto Nacional de Investigaciones Nucleares for the use of their computing facilities. One of us (JK) thanks the Deutscher Akademischer Austauschdienst (DAAD) of Germany for a visiting fellowship to the University of Konstanz where part of this work was done.

References

Abia C, Domínguez I, Straniero O, Limongi M, Chieffi A and Isern J 2001 *ApJ* **557**, 126
Alexander D R and Ferguson J W 1994 *ApJ* **437** 879
Angulo C, Arnould M, Rayer M *et al* 1999 *Nucl. Phys.* A **656** 3
Arnould M, Goriely S and Jorissen A 1999 *A&A* **347** 572
Bahena D 2006 *PhD Thesis* Univerzita Karlova v Praze
Bahena D, Klapp J and Dehnen H 2007 *Rev. Mex. Fís* **53**, 48
Bahena D and Klapp J 2010 *ApSS* **327**, 219
Berrington K A, Burke P G, Butler K *et al* 1987 *J. Phys. B: Al. Mol. Phys.* **20** 6369
Bromm V, Kudritzki R P and Loeb A 2001 *ApJ* **552** 464

Buchmann L 1996 *ApJL* **468** 127
Cassisi S 2005 [arXiv: astro-ph/0506161
Caughlan G R and Fowler 1988 *At. Data Nucl. Data Tables* **40** 283
Caughlan G R, Fowler W A, Harris M J and Zimmerman B A 1985 *At. Data Nucl. Data Tables* **32** 197
Charbonnel C, Meynet G, Maeder A, Schaller G and Schaerer D 1993 *A&ASS* **101** 415
Christy R 1966 *ApJ* **144** 108
Claret A 1995 *A&AS* **109** 441
Claret A 2004 *A&AS* **424** 919
Claret A 2005 *A&A* **440** 647
Claret A and Giménez A 1991 *A&AS* **87** 507
Claret A and Giménez A 1992 *A&AS* **96** 255
Cox A N and Steward J N 1965 *ApJS* **11** 22
Cox A N and Steward J N 1970a *ApJS* **19** 243
Cox A N and Steward J N 1970b *ApJS* **19** 261
Eggleton P P, Faulkner J and Flannery B P 1973 *A&A* **123**
Fowler W A, Caughlan G and Zimmerman B A 1967 *ApJ* **5** 525
Fowler W A, Caughlan G and Zimmerman B A 1975 *ApJ* **13** 69
Fryer C L, Woosley S E and Heger A 2001 *ApJ* **550** 372
Formicola A *et al* 2004 *Phys. Lett.* B **591** 61
Graboske H C, DeWitt H L, Grossmann A S and Cooper M S 1973*ApJ* **181** 457
Harris M J, Fowler W A, Caughlan G and Zimmerman B A 1983 *ARA&A* **21** 165
Heger A and Woosley S E 2002 *ApJ* **567** 532
Heger A, Woosley S E, Rauscher T, Hoffman R D and Boyes M M 2002 *New Astron. Rev.* **46** 463
Hubbard W B and Lampe M 1969 *ApJS* **18** 297
Huebner W F, Merts A L, Magee N H and Argo M F 1977 *Los Alamos Scientific Report*, LA-6760-M
Iglesias C A and Rogers F J 1991 *ApJS* **371** 408
Iglesias C A and Rogers F J 1993 *ApJS* **412** 752
Iglesias C A and Rogers F J 1996 *ApJS* **464** 943
Iglesias C A, Rogers F J and Wilson B G 1992 *ApJ* **397** 717
Klapp J 1982 *PhD Thesis*, University of Oxford
Klapp J 1983 *ApSS* **93** 313
Klapp J 1984 *ApSS* **106** 215
Klapp J and Bahena D 2008 *Rev. Mex. Fís S* **54** 74
Klapp J, Bahena D, Corona-Galindo M G and Dehnen H 2005 in *Gravitation and Cosmology*, eds. A Macías, C. Lämmerzahl and D Nuñez, Melville N Y, AIP Conf. Proc. **758** 153
Klapp J, Bahena D, Corona-Galindo M G, Dehnen H and Galindo S 2006 *Rev. Mex Fís*, **52** 50
Kurucz R L 1991 in *Stellar Atmospheres: Beyond Classical Models*, NATO ASI Series C, 341
Landré V, Prandtzos N, Aguer P *et al* 1990 *A&A* **240** 85
Maeder A 1990 *A&AS* **84** 139
Marigo P, Girardi L, Chiosi C and Wood P R 2001 *A&A* **347** 152
Rogers F J and Iglesias C A 1992 *ApJS* **79** 507
Rolfs C E and Rodney W S 1988 *Cauldrons in the Cosmos*, University of Chicago
Runkle R C 2003 *PhD Thesis*, University of North Carolina
Seaton M J 1987 *J. Phys. B: At. Mol. Phys.* **20** 6363
Schaerer D, Meynet G, Maeder A and Schaller G 1993a *A&AS* **98** 523
Schaerer D, Charbonnel C, Meynet G, Maeder A and Schaller G 1993b *A&AS* **102** 339
Schaller G, Schaerer D, Meynet G and Maeder A 1992 *A&AS* **96** 269
Siess L, Livio M and Lattanzio J 2002 *ApJ* **570** 329
Stothers R and Chin C W 1991 *ApJ* **381** L67
Tumlinson J, Shull J M and Venkatesan A 2003 *ApJ* **5287** L65
Umeda H, Nomoto K and Nakamura T 2000, *In The First Stars: Proceedings of the MPA/ESO Workshop Held at Garching, Germany, 4-6 August 1999, ESO Astrophysics Symposia.* Edited by A. Weiss, T.G. Abel, and V. Hill. Springer-Verlag, p. 150
Weaver T A and Woosley S E 1993 *Phys. Rep.* **227** 65
Woosley S E and Weaver T A 1995 *ApJS* **101** 181
Woosley S E, Langer N and Weaver T A 1993 *ApJ* **411** 823
Weiss A, Keady J J and Magee N H 1990 *Atomic and Nuclear Data Tables* **45** 209

Journal of Physics: Conference Series **239** (2010) 012005 doi:10.1088/1742-6596/239/1/012005

Phase transitions in algebraic cluster models

H Yépez-Martínez[1], L Parra Rodríguez[2], P O Hess[2], J Cseh[3] and G Lévai[3]

[1] Universidad Autónoma de la Ciudad de México, Prolongación San Isidro 151, Col. San Lorenzo Tezonco, Del. Iztapalapa, 09790 México, D.F., Mexico
[2] Instituto de Ciencias Nucleares, UNAM, A.P. 70-543, 04510 México, D.F., Mexico
[3] Institute of Nuclear Research of the Hungarian Academy of Sciences (ATOMKI), Debrecen, Pf. 51, Hungary-4001

E-mail: hess@nucleares.unam.mx

Abstract. Two algebraic cluster models are studied from the point of view of phase transitions: one in which the Pauli exclusion principle is taken into account and one in which it isn't. A third-order interaction is introduced to avoid instabilities in the model spectra, which is generally not taken into account. It is shown that both first- and second-order phase transitions occur. The ^{20}Ne$\rightarrow^{16}$O$+\alpha$ system is considered as an example. Without taking into account the Pauli exclusion principle the transition from the $SU(3)$ to the $SO(4)$ dynamical symmetry is of first or second order, depending on the strength of the quadrupole-quadrupole interaction. It is shown that the inclusion of the Pauli-principle can be simulated by higher-order interactions when the model space is not truncated.

1. Introduction

The study of phase transitions enjoys a substantial interest in algebraic models of nuclear structure. The first examples [1, 2] were related to the Interaction Boson Approximation (IBA) [3] and the vibron model [4, 5]. In these approaches a coherent state is constructed in terms of collective variables and a semi-classical potential is defined as the expectation value of the Hamiltonian with respect to the coherent state. Phase changes are identified with discontinuities in the derivatives of the potential with respect to specific parameters taken at the potential minima. An important conjucture was that phases are determined by quasi-dynamical symmetries. This approximate symmetry is also denoted as an effective symmetry and its mathematically sound definition, called embedded symmetry [6] has been applied to the shell model of the nucleus in [7]. The role of the quasi-dynamical symmetry in relation with the phase-transition was discussed in [8, 9, 10].

Here we discuss phase transitions within algebraic cluster models, in which the relative motion is described by the vibron model. It was found that the corresponding phase transition is of second order [11]. In [12] phase transitions in $U(n)$ algebraic models were discussed using interaction terms up to second order and the existence of a second-order phase transition was confirmed. In [13] a numerical study was performed within a realistic cluster system and it was argued (without proof) that the transition might be of first order.

Here we revisit phase transitions within a particular set of cluster models. One is the Phenomenological Algebraic Cluster Model (PACM), while the other is the Semimicroscopic Algebraic Cluster Model (SACM) [14, 15]. The first set does not obey the Pauli exclusion principle, while the latter does. We apply the coherent state method to investigate phase transitions and their order. We consider two spherical clusters in order to discuss the following points: i) What changes when we implement the Pauli exclusion principle? ii) Will the order of phase transition change when third-order interactions are introduced to stabilize the spectrum? We will show that there are first- and second-order phase transitions in the PACM. iii) We will show that when the Pauli exclusion principle is taken into account, it corresponds to introducing higher-order interactions (in the PACM language). This changes significantly the structure of phase transitions and may involve a critical point.

In Section 2 we resume the cluster models considered including the Hamiltonian, which will be the same for the two groups of cluster models. This section also deals with the determination of the order of phase transitions using standard methods [16, 17]. In Section 3 the system ^{20}Ne$\rightarrow^{16}$O$+\alpha$ is considered as a particular example, while in Section 4 conclusions are drawn.

2. The cluster models and phase transitions

2.1. The model space

We consider two different kinds of cluster model: the Hamiltonian in both is exactly the same and they differ in their model space. In the PACM the model space includes states which are forbidden by the Pauli exclusion principle, while within the SACM the Pauli exclusion principle is taken into account. The basic degrees of freedom are dipole bosons $\boldsymbol{\pi}_m^\dagger$ and scalar bosons $\boldsymbol{\sigma}^\dagger$. The latter are introduced in order to generate a cutoff N, such that the total number of bosons, $n_\pi + n_\sigma = N$ is kept constant. In the SACM the relevant group chain defining the basis is $SU_{C_1}(3) \otimes SU_{C_2}(3) \otimes SU_R(3) \supset SU_C(3) \otimes SU_R(3) \supset SU(3) \supset SO(3) \supset SO(2)$, from the viewpoint of the space degrees of freedom. The individual clusters are described by the irreducible representation (irrep) of $SU_{C_k}(3)$ and are denoted by (λ_k, μ_k). The two cluster irreps are coupled to the irrep (λ_C, μ_C) of $SU_{C_k}(3)$. The relative motion is described by the $SU_R(3)$ group with its irrep $(n_\pi, 0)$, where n_π is the number of oscillator quanta in the relative motion. (λ, μ) is the total irrep of $SU(3)$. L and M are the angular momentum and its projection of the $SO(3)$ and $SO(2)$ group, respectively, and there is a multiplicity index from $SU(3) \supset SO(3)$ in order to distinguish multiple occurrences of a given L in (λ, μ). The individual clusters are described in terms of Elliott's $SU(3)$ model [18]. All possible $SU(3)$ irreps (λ, μ) are determined. Within the PACM no restriction is applied and the number of π bosons starts from zero. Within the SACM the list of the $SU(3)$ irreps is compared to the list of possible irreps within the $SU(3)$ shell model and only those appearing in the latter list are taken into account. (A similar selection is applied also for the irreps of the $U^{ST}(4)$ spin-isospin group.) As a consequence, there is a minimal number of π bosons, denoted by n_0. In what follows only the case of two spherical clusters is considered, which implies $(\lambda_\alpha, \mu_\alpha) = (0,0)$, $\alpha = 1, 2$ and $(\lambda, \mu) = (n_\pi, 0)$.

2.2. The Hamiltonian

The Hamiltonian, for the case of two spherical clusters is given by

$$\boldsymbol{H} = x\boldsymbol{H}_{SU(3)} + (1-x)\left(\boldsymbol{H}_{O(4)} - \frac{1}{2}\right) \quad , \tag{1}$$

where the term $-\frac{1}{2}$ is introduced for convenience. Here

$$\begin{aligned} \boldsymbol{H}_{SU(3)} &= \hbar\omega \boldsymbol{n}_\pi + y(a - b\Delta\boldsymbol{n}_\pi)\boldsymbol{n}_\pi(\boldsymbol{n}_\pi + 3) + \gamma \boldsymbol{L}^2 \\ \boldsymbol{H}_{O(4)} &= \frac{c}{4}[(\boldsymbol{\pi}^\dagger \cdot \boldsymbol{\pi}^\dagger) - (\boldsymbol{\sigma}^\dagger)^2][(\boldsymbol{\pi} \cdot \boldsymbol{\pi}) - (\boldsymbol{\sigma})^2] + \gamma \boldsymbol{L}^2 \quad , \end{aligned} \tag{2}$$

with $\Delta\boldsymbol{n}_\pi = \boldsymbol{n}_\pi - n_0$, n_0 being the minimal number of quanta, which, in the PACM is equal to zero, while in the SACM it differs from zero. For the case of two spherical clusters the second-order Casimir operator of $SU(3)$ is just given by $\boldsymbol{n}_\pi(\boldsymbol{n}_\pi + 3)$. The variable y gives the intensity

Journal of Physics: Conference Series **239** (2010) 012005 doi:10.1088/1742-6596/239/1/012005

of the quadrupole-quadrupole interaction. It is used as parameter which defines the strength of the quadrupole-quadrupole interaction and, therefore, varies only from $y = 1$ (full interaction) to $y = 0$ (no quadrupole-quadrupole interaction). $H_{SU(3)}$ contains a third-order interaction $\Delta \boldsymbol{n}_\pi \boldsymbol{n}_\pi^2$, too. This term is of utmost importance in stabilizing the spectrum. In order to get a stable spectrum (with no high n_π states getting below the ground state), the parameter a has to be negative (corresponding to an attractive quadrupole-quadrupole interaction) and the parameter b has to have the same sign as a. Without this term states with large enough n_π would come below the ground state. The justification of the third-order term is given in [19] within the context of the symplectic model.

The proposed correction to the strength of quadrupole-quadrupole interaction is the simplest possible one. In principle it could be replaced by the third-order Casimir operator of $SU(3)$ too.

2.3. Coherent state and the mapping to a semi-classical potential

In order to map to a semi-classical potential, a coherent state is introduced as [20],

$$\mid \alpha > \;=\; \mathcal{N}_{N,n_0} \frac{N!}{(N+n_0)!} \frac{d^{n_0}}{d\gamma^{n_0}} \left[\boldsymbol{\sigma}^\dagger + \gamma \left(\boldsymbol{\alpha} \cdot \boldsymbol{\pi}^\dagger \right) \right]^{N+n_0} \mid 0 > \tag{3}$$

where the γ parameter has to be set to 1 after the differentiation and the normalization factor $\mathcal{N}_{N,n_0}$ is given in [20]. For convenience here we redefined the total number of relative oscillation quanta by $(N + n_0)$. The coherent state for the PACM is obtained with $n_0 = 0$. The only variable which will play a role is the absolute value of $\boldsymbol{\alpha}_m$, denoted by α, with $\alpha^2 = \sum_m \alpha_m \alpha^m$.

The geometrical potential is obtained by calculating the expectation value of $\boldsymbol{H}$ with respect to the coherent state using $\beta^2 = \alpha^2/(1+\alpha^2)$. Within the PACM the result is particularly simple:

$$\begin{aligned} \langle \boldsymbol{a} | \boldsymbol{H} | \boldsymbol{a} \rangle \;=\; V(\beta) \;=\; & N\beta^2 \left[x \left(\hbar\omega + 4y \left(a - b \right) \right) - (1-x)\, c \,(N-1) \right] \\ & + N\,(N-1)\,\beta^4 \left[xy \left(a - 6\, b \right) + (1-x)\, c \right] - N\,(N-1)\,(N-2)\,\beta^6 xyb \quad . \end{aligned} \tag{4}$$

This potential can be reparametrized as

$$V \;=\; N(N-1)(N-2)(-bx) \left\{ A\beta^2 - B\beta^4 + \beta^6 \right\} \quad . \tag{5}$$

Finally, it suffices to investigate the effective potential $\widetilde{V} = \left\{ A\beta^2 - B\beta^4 + \beta^6 \right\}$, which depends on the variable $\beta \in [0, 1]$.

In the SACM the expression gets more involved. For example, the expectation value of $\boldsymbol{n}_\pi$ in the PACM is given by $N\alpha^2/(1 + \alpha^2)$, while within the SACM we obtain

$$\langle \boldsymbol{n}_\pi \rangle \;=\; (N + n_0)\alpha^2 \frac{F_{11}(\alpha^2)}{F_{00}(\alpha^2)} \quad . \tag{6}$$

The F_{pq} function is a polynomial in α^2 of the order $N + n_0 - max(p, q)$. Similar formulas are obtained for the other interaction terms. This expression makes it extremely difficult to determine analytically the first and second derivatives of the potential with respect to the parameters of the model, taken at generally more than one minimum of the potential. As we shall see below, there is still a possibility to determine them.

There is one important observation: Inspecting Eq. (6) and noting that within the PACM $\langle \boldsymbol{n}_\pi \rangle = N\frac{\alpha^2}{1+\alpha^2}$ holds, one can realize that the inclusion of the Pauli exclusion principle corresponds to the introduction of higher-order interactions of the type $\boldsymbol{n}_\pi^k$.

Another observation is related to the N-dependence of the potential. In principle, one can renormalize the expression such that no N-dependence appears at the end (for example, see [21]). Due to lack of space, we do not apply this renormalization and refer to a later publication.

By inspecting the expressions for the SACM we immediately note some important differences to PACM: the expectation value of the π number operator is proportional to $\alpha^2/(1+\alpha^2)$, while within the SACM we get a complicated function in α. This function can also be obtained within the PACM when higher-order interaction terms are included. Thus the ratios F_{pq}/F_{00} simulate higher-order interactions within the PACM. This already indicates that the phase transitions can be of order higher that two.

2.4. Phase transitions

Phase transitions are investigated using the following steps recommended by [16, 17, 22, 23].

a) First the minima of the Potential Energy Surface (PES) in the space of the collective variables α_m is determined. Now there is only one relevant variable α, because the distance vector between the two clusters can always be aligned along the intrinsic z-axis. Studying the extrema via $\frac{d\widetilde{V}}{d\alpha}=0$ determines the potential at the minima $V(\bar{\alpha}_i)$ and their location $\alpha = \bar{\alpha}_i$ ($i = 1, 2, ..,$ which numbers the different minima). Note that the index i now does not refer to a component of α but to the value of α at a specific minimum. The $\bar{\alpha}_i = \bar{\alpha}_i(x)$ are functions of the interaction parameters $\vec{p} = (p_k)$, with $k = 1, 2, ..., n_k$ and n_k being considered as control parameters. In our case there are two, namely x and y. The y parameter serves only in fixing the strength of the quadrupole-quadrupole interaction (parameters a and b).

b) Once the extrema are obtained, one determines at each minimum the derivatives $\frac{\partial^n \widetilde{V}(\bar{\alpha}_i(\vec{p})}{\partial p_k^n})$ of the potential with respect to the control parameters: in most cases the first and second derivatives are sufficient. In this contribution the free parameters are x and y, although the interaction parameters can, in principle, be varied, too. Nevertheless, here we shall fix the interaction constants to the spectrum of realistic systems at the points of exact dynamical symmetries and then vary the x and y variables only.

c) The places of phase transitions are determined identifying the borders in the parameter space where at least two minima are at equal energy, i.e., $\widetilde{V}(\bar{\alpha}_{i_1}) = \widetilde{V}(\bar{\alpha}_{i_2})$ for some index values i_1 and i_2 of the minima. This provides us with a relation $f(p_1, ..., p_{n_p}) = 0$ between the parameters, which allows, in principle, to express one parameter in terms of the others.

d) The phase transition is of order m, when the m'th derivative of the potential is discontinuous at the point of the phase transition, i.e.,

$$\frac{\partial^n \widetilde{V}(\bar{\alpha}_{i_1})}{\partial p_k^n} = \frac{\partial^n \widetilde{V}(\bar{\alpha}_{i_2})}{\partial p_k^n} \quad , \quad \text{for } n < m \quad , \quad \frac{\partial^m \widetilde{V}(\bar{\alpha}_{i_1})}{\partial p_k^m} \neq \frac{\partial^m \widetilde{V}(\bar{\alpha}_{i_2})}{\partial p_k^m} \quad . \tag{7}$$

The derivative of $\widetilde{V}(\bar{\alpha}_i)$ with respect to a parameter has to be understood as a derivative of $\widetilde{V}(\alpha)$ with respect to this parameter and after that α is set to $\bar{\alpha}_i$. Note that this procedure of determining the order of a phase transition is quite general and does not depend on identifying the phase by a dynamical symmetry.

The derivatives of the potential are easily determined within the PACM, however, within the SACM this is not so any more. We consider the particular potential structure, valid only for the particular case discussed in this contribution,

$$\widetilde{V}(\alpha) = x\widetilde{V}_1(\alpha) + (1-x)\widetilde{V}_2(\alpha) \quad . \tag{8}$$

implying the division into a $SU(3)$ and $SO(4)$ part given in Eq. (1). The difference to Eq. (5) consists in the division of a simple factor, or just the expectation value calculated within the curly bracket. The minima of $\widetilde{V}$ are obtained using algebraic and numerical routines. Taking the values of the potential $\widetilde{V}(\alpha)$ at the minima, the derivative with respect to x is given by

$$\frac{d\widetilde{V}(\bar{\alpha}_i)}{dx} = \frac{\partial \widetilde{V}(\bar{\alpha}_i)}{\partial x} + \frac{\partial \widetilde{V}(\bar{\alpha}_i)}{\partial \bar{\alpha}_i}\frac{\partial \bar{\alpha}_i}{\partial x} = \frac{\partial \widetilde{V}(\bar{\alpha}_i)}{\partial x} = \widetilde{V}_1(\bar{\alpha}_i) - \widetilde{V}_2(\bar{\alpha}_i) \tag{9}$$

with $i = 1$ or 2. We exploited the fact that the first derivatives of the potential at a minimum are equal to zero. These expressions have to be calculated at the point of phase transition.

In case $\widetilde{V}(\bar{\alpha}_i)$ turns out to be identically zero or constant for a given i (as will be the case in our example for $\bar{\alpha}_1$, which will be related to the spherical minimium where $V = 0$), then by construction all higher derivatives also give zero. Therefore, in what follows, higher derivatives are only computed when $\widetilde{V}(\bar{\alpha}_i)$ is not a constant.

There are now two possibilities. i) The two minima are separated, i.e., at the point of phase transition the deformed minimum is at a value $\bar{\alpha}_2$ different from zero, while the spherical minimum is at $\bar{\alpha}_1 = 0$. *ii)* The two minima coincide at the point of phase transition, i.e., $\bar{\alpha}_1 = \bar{\alpha}_2$ at $x = x_0$, with x_0 denoting the value x at which the phase transition takes place.

For case i) the expression in (9) is in general different for the two minima, except for a rare coincidence when $\widetilde{V}_1(\bar{\alpha}_1) - \widetilde{V}_2(\bar{\alpha}_1) = \widetilde{V}_1(\bar{\alpha}_2) - \widetilde{V}_2(\bar{\alpha}_2)$ for $\bar{\alpha}_1 \neq \bar{\alpha}_2$. Thus we can conclude that in this case the first derivatives at the point of phase transition are different, taken at α_1 in one case and at α_2 in the other case. Therefore, the phase transition is of order one.

In the second case the second derivative also has to be calculated:

$$\frac{d^2\widetilde{V}(\bar{\alpha}_i)}{dx^2} = \left[\frac{\partial\widetilde{V}_1(\bar{\alpha}_i)}{\partial\bar{\alpha}_i} - \frac{\partial\widetilde{V}_2(\bar{\alpha}_i)}{\partial\bar{\alpha}_i}\right]\frac{\partial\bar{\alpha}_i}{\partial x} \quad . \tag{10}$$

For the case when the positions of the two minima coincide, the expression in the squared parenthesis for $\alpha = \bar{\alpha}_1$ and $\alpha = \bar{\alpha}_2$ are equal. However, the derivatives of $\bar{\alpha}_i$ with respect to x are, in general, different, except for the coincidence, when the factor of the derivative of $\bar{\alpha}_i$ is the same for $\bar{\alpha}_1$ and $\bar{\alpha}_2$. Thus we can conclude that the phase transition is of second order here. Of course, the numerical values have to be checked in order to exclude the rare coincidence. In order to do that higher derivatives have to be determined, which is cumbersome in the SACM.

This new consideration makes it manageable to determine the order of a phase transition within the SACM. All we have to do is to determine if the two minima are separated at the point of phase transition (first-order) or merge at the point of phase transition (second-order).

The general phase diagram has the following structure (see left panel Fig. 1): For $A > 0$ and $B \leq 0$ only a spherical minimum exists, while for $A < 0$ only a deformed minimum exits. For $B > 0$ and $0 \leq A \leq \frac{B^2}{3}$ a spherical and a deformed minimum co-exist. For $A > \frac{B^2}{4}$ the gobal minium is the spherical one, while for $A < \frac{B^2}{4}$ it is the defomed one. At $A = \frac{B^2}{4}$ a phase transition takes place which is of first order. Thus the phase diagram shows a much richer structure, including first- and second-order phase transitions.

3. A particular application: $^{20}Ne \rightarrow ^{16}O + \alpha$

In this section we consider a system that has the same model space as ^{20}Ne$\rightarrow^{16}$O$+\alpha$. In reality we do not consider the physical ^{20}Ne, because this is given by a fixed value of x and y adjusted to the physical spectrum. Rather, we describe a hypothetical cluster system with the same $SU(3)$ representation as ^{20}Ne, changing its structure going from $x = 1$, the $SU(3)$ dynamical symmetry limit, to $x = 0$, the $SO(4)$ dynamical symmetry limit. In the first step we considered the PACM only. We adjusted the interaction parameters of the model such that the energy levels, both within the $SU(3)$ and $SO(4)$ dynamical symmetry limit exhibit the same scale and resemble the physical spectrum in the $SU(3)$ dynamical symmetry limit. As a result we obtained for the parameters of (4) the values $a = -2.311$ MeV, $b = -0.17$ MeV, $c = 0.335$ MeV and we used $n_0 = 8$, $N = 20$ and $(\lambda, \mu) = (8, 0)$. In Fig. 1 the corresponding phase diagram is plotted in the left panel.

The horizontal axis gives x and shows the transition from the $SO(4)$ dynamical symmetry for $x = 0$ towards the $SU(3)$ dynamical symmetry at $x = 1$. The vertical axis shows the value y,

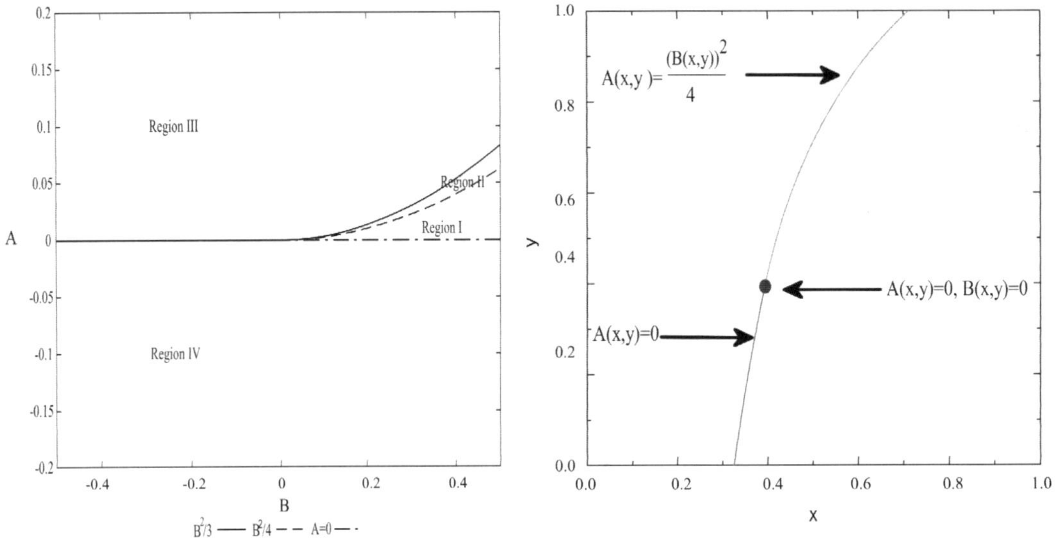

Figure 1. The left panel shows the phase space diagram as a function of the two parameters A and B. See the text for explanations. The right panel shows the phase diagram for the system ^{20}Ne $\rightarrow$ ^{16}O + α. Note that there is a first-order phase transition line starting from the upper right side of the curve until reaching the dot. To the left of the dot the phase transition is of second order.

which is a measure of the strength of the quadrupole-quadrupole interaction. One has to read for a fixed y the horizontal axis x.

Our first observation is that removing the three-body interaction the results of Ref. [12] are recovered, i.e. a second-order phase transition occurs.

The line of phase transition is also plotted for the case when the three-body interaction is included in general. The line is divided into two sections, the division indicated by a dot. The upper right part of the curve indicates a first-order phase transition, while the lower left section indicates a second-order phase transition. For $y = 1$ we find a first-order phase transition. When y is lowered, from one point on, when the quadrupole-quadrupole interaction is sufficiently small, the phase transition turns over to second-order. Thus, we found an example when there exists not only a second-order phase transition, but also a first-order one. A small strength of the quadrupole-quadrupole interaction corresponds to unharmonic corrections in the interaction and reproduces the known results. The new result is that when the interaction is stongly unharmonic (including a cubic term) the transition turns over to a first-order one.

We repeated the investigation for the SACM, when the Pauli exclusion principle is taken into account. Further above we noted that this introduces more involved expressions, corresponding to higher-order interactions within the PACM. The same interaction parameters are used and only x and y are varied. We used the method explained in the last section. For the case of ^{20}Ne, we found that the phase transition changes dramatically: For $y = 1$ the phase transition between the $SU(3)$ and $SO(4)$ dynamical symmetries vanishes! This is because the potential at $y = 1$ has a deformed minimum for all values of x, so no spherical minimum appears! As a consequence, no structural change happens and, thus no phase transition. At $y = 0$ the situation changes. Now at $x = 1$ there is a spherical minimum. In between $x = 0$ and $x = 1$ this spherical minimum turns over into a deformed one. Since the two minima merge at the point of phase transition, it is of order two.

We have not done yet he analysis for y values between 0 and 1. However, one can safely make the following conjecture: somewhere between $y = 0$ and 1 the phase transition ceases to exist, implying a critical point. All this indicates that the structure of phase transitions exhibits a more dramatic structure than in the previous considerations, including the existence of a critical point.

4. Conclusions

We investigated the phase transition properties within algebraic cluster models, which are divided into two classes called PACM and SACM. In the first class the Pauli exclusion principle is not taken into account, while in the second it is. In order to avoid instabilities we had to include a third-order interaction in the Hamiltonian. The results are the following:
i) The same results are obtained when only up to two-body interactions are considered [12] and the Pauli exclusion principle is neglected. The differences appear with the inclusion of higher-order interactions and/or when the Pauli exclusion principle is taken into account.
ii) When a three-body interaction is included within the PACM, the structure of the phase diagram changes. It now also includes first-order phase transitions.
iii) When the Pauli exclusion principle is taken into account (SACM), from the point of view of the PACM it looks equivalent to introducing complicated higher-order interactions. The structure of phase transitions changes significantly, allowing also the appearance of critical points.

Acknowledgments

The authors acknowledge support from CONACyT-MTA, DGAPA and OTKA (grant No K72357).

References

[1] Leviatan A and Kirson M W 1988 *Ann. Phys. (N.Y.)* **188** 142
[2] Roosmalen O S 1982 *Algebraic Description of Nuclear and Molecular Rotation-Vibration Spectra* (Ph.D. Thesis, Groningen, The Netherlands)
[3] Iachello F and Arima A 1987 *The Interacting Boson Model* (Cambridge University Press, Cambridge)
[4] Iachello F 1981 *Phys. Rev.* C **23** 2778
[5] Iachello F and Levine R D 1982 *J. Chem. Phys.* **77** 3046
[6] Rochford P and Rowe D J 1988 *Phys. Lett.* B **210** 5;
Rowe D J, Rochford P and Repka J 1988 *J. Math. Phys.* **29** 572
[7] Jarrio M, Wood J L and Rowe D J 1991 *Nucl. Phys.* A **528** 409
[8] Rowe D J 2004 *Nucl. Phys.* A **475** 47
[9] Cseh J, Darai J, Yépez-Martínez H and Hess P O 2008 *Int. J. Mod. Phys.* E **17** 2296
[10] Cseh J 2010 *J. Phys.: Conf. Series* **205** 012021
[11] Zhang Y, Hou Z F, Chen H, Wei H and Liu Y X 2008 *Phys. Rev.* C **78** 024314
[12] Cejnar P and Iachello F 2007 *J. Phys. G: Nucl. Part. Phys.* **40** 581
[13] Yépez-Martínez H, Cseh J and Hess P O 2006 *Phys. Rev.* C **74** 024319
[14] Cseh J 1992 *Phys. Lett.* B **281** 173
[15] Cseh J and Lévai G 1994 *Ann. Phys. (N.Y.)* **230** 165
[16] Gilmore R 1979 *J. Math. Phys.* **20** 891
[17] Gilmore R 1981 *Catastrophe Theory for Scientists and Engineers* (Wiley, New York)
[18] Elliott J P 1958 *Proc. Roy. Soc.* A **245** 562
[19] Rosensteel G and Draayer J P 1985 *Nucl. Phys.* A **436** 445
[20] Hess P O, Lévai G and Cseh J 1996 *Phys. Rev.* C **54** 2345
[21] Cejnar P and Jolie J 2009 *Prog. Part. Nucl. Phys.* **62** 210
[22] López-Morena E and Castaños O 1996 *Phys. Rev.* C **54** 2374
[23] Castaños O, López-Peña R, Hirsch J G and López-Moreno E 2006 *Phys. Rev.* B **74** 104118

On the hyperdeformed state of the ^{36}Ar nucleus

J Cseh[1], J Darai[2], N V Antonenko[3], G G Adamian[3], A Algora[1,4], P O Hess[5] and A Lépine-Szily[6]

[1] Institute for Nuclear Research, Hung. Acad. Sci., Debrecen, POB 51, Hungary-4001
[2] Inst. of Experimental Phys., Univ. of Debrecen, Debrecen, POB 105, Hungary-4010
[3] Bogoliubov Laboratory, Joint Institute for Nuclear Research, 141980 Dubna, Russia
[4] IFIC (CSIC-Univ. de Valencia), A. C. 22085, E 46071 Valencia, Spain
[5] Instituto de Ciencias Nucleares, UNAM, A.P. 70-543, 04510 Mexico, D.F., Mexico
[6] Instituto de Fisica, Univ. de São Paulo, CP 66318, 05389-970 São Paulo, Brazil

E-mail: cseh@atomki.hu

Abstract. The hyperdeformed state of the ^{36}Ar nucleus is discussed, including its prediction from alpha-cluster and binary cluster studies, via its experimental verification from heavy-ion resonances up to its derivation from Nilsson-model calculations. Its possible clusterizations are also investigated both from the microscopic structure and from the energetic preference viewpoint.

1. Introduction

The investigation of extreme, e.g. superdeformed (SD) and hyperdeformed (HD) nuclear shapes is one of the exciting topics in the present day nuclear structure research. Much attention is focused on the $N = Z$ nuclei, in which the proton-neutron interactions, clusterization, etc. play especially important role. In these nuclei even more exotic shapes, like e.g. linear alpha-chains have been predicted. In spite of the considerable experimental and theoretical efforts, however, only very limited convincing evidence is available [1].

Apart from some very simple and well-known cluster states, like e.g. the ground state of the ^{8}Be nucleus, the superdeformed (SD) states of these self-conjugate nuclei have been observed experimentally only during the last decade. The SD band of ^{36}Ar was reported for the first time in [2], populated in an ^{24}Mg(^{20}Ne,2α)^{36}Ar reaction, and detected by multiple (gamma-ray and charged-particle) coincidence techniques. Following the experimental observation a lot of theoretical effort was concentrated on this state, therefore, we have a fairly good theoretical understanding of it. According to the joint conclusion of large-scale shell model, Nilsson model, different Hartree-Fock, and various cluster-model calculations this state has 4 nucleon excitation (for a brief recent review and references see [3]).

The question of the hyperdeformed (HD) state of these nuclei was so far mainly a territory for theoretical predictions. For ^{36}Ar it was first predicted by an alpha-cluster model calculation [4], giving the moment of inertia of the state. Then based on binary cluster studies it was predicted that it can be populated predominantly in ^{24}Mg+^{12}C, and ^{20}Ne+^{16}O reactions [5]. A recent study on the ^{24}Mg+^{12}C elastic scattering, completed with some available ^{20}Ne+^{16}O reaction data seems to verify this prediction [6]. The moment of inertia is in good agreement with the prediction of the alpha-cluster model. In [3, 5] the possible binary clusterizations have been

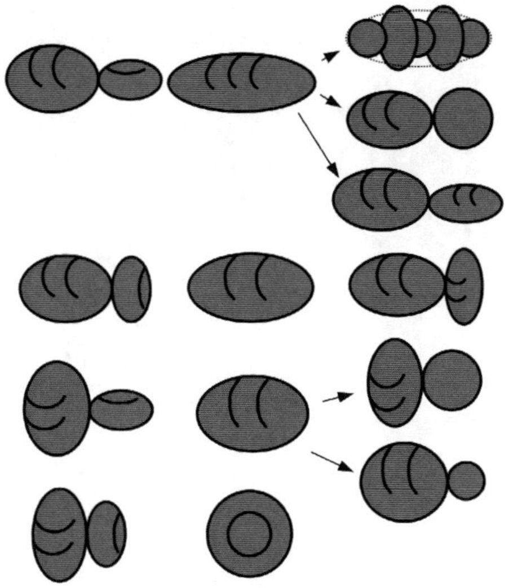

Figure 1. The shape isomers of the ^{36}Ar nucleus from Nilsson-model calculations, and their amalgamation from two clusters. The central part shows the shell-model results for the ground (at the bottom), superdeformed, triaxial (HD_b), and hyperdeformed (at the top) states. The left column corresponds to the ^{24}Mg+^{12}C clusterization. The right side illustrates the ^{32}S+^{4}He, ^{20}Ne+^{16}O, ^{28}Si(prolate)+^{8}Be, ^{28}Si(prolate)+^{8}Be, ^{20}Ne+^{16}O configurations (from the bottom), respectively. At the upper most line also the alpha-clusterization of the HD state is shown from the work [4], where the contour corresponds to the HD state of the Nilsson-model.

extensively discussed from the microscopic structure viewpoint, and here we present detailed double-folding calculations for the energetic preference of different configurations. Since the theoretical predictions on the HD state came from cluster models, it is highly desirable to study, whether or not this state shows up also in the shell model. A recent calculation within the framework of the Nilsson-model resulted in a HD state, which has not only the same moment of inertia, as obtained from alpha-cluster model, and observed experimentally, but it has also the same $U(3)$ symmetry of the state [3], as obtained in [4].

In what follows we will discuss these aspects more in detail.

2. Theoretical predictions from cluster studies

In [4], and later on in [7] unconstrained Bloch-Brink alpha-cluster model calculations have been carried out. A hyperdeformed state with ratios of major axes 3:1:1 was found [4], and the corresponding shell-model configuration is given in Table 1 of [7]. (Please, note that a $[400]^4$ part is missing there due to printing error [8].) The geometrical arrangement of the 9 α-particles is a $\alpha - C - \alpha - C - \alpha$ linear chain, where C stands for an oblate ^{12}C nucleus, i.e. 3 α's in a triangular arrangement, and the carbon nuclei line up their symmetry axes with the molecular axis, as shown in Fig. 1. The corresponding shell-model configuration indicate a [48,8,8] $U(3)$ symmetry for this state. The moment of inertia (corresponding to this symmetry) turns out to be 4.2 in units of 10^5 fm^2 MeV/c^2.

The allowed binary clusterizations of the superdeformed and hyperdeformed states in ^{36}Ar (together with those of the ground state) were investigated systematically in [5]. The clusters were supposed to be in their intrinsic ground state, with spherical or deformed (even triaxial) shapes, with no simplifying assumption for their relative orientation and all the stable isotopes were considered as possible clusters. The problem was investigated both from the viewpoint

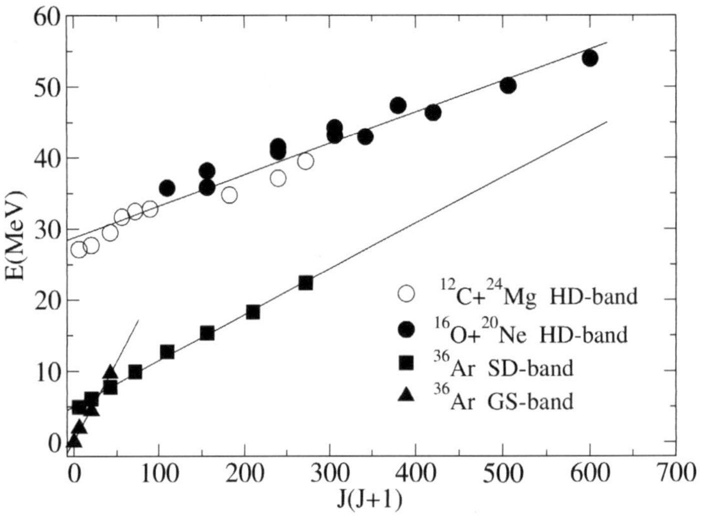

Figure 2. Rotational bands in the ^{36}Ar nucleus built on the ground, superdeformed and hyperdeformed states. The symbols are experimental results, and the lines are fitted to them.

of the microscopic structure and from the angle of energetic preference, calculated in terms of the criterion of maximum stability (combined with the no-dipole constraint) [9]. The structural aspect was treated by the combined application of the $U(3)$ selection rule and Harvey's prescription (for a general discussion see [10]).

In [5] the superdeformed and hyperdeformed states were defined in two different ways: on the one side by their quadrupole deformation (from the experiment for the superdeformed state and by the usual $\beta \approx 0.86$ value for the hyperdeformed state, noted there as SD(b) and HD(b) states, respectively), or by corresponding simple shell model configurations noted as SD(a) and HD(a).

This study revealed that the ground state allows more asymmetric, while the hyperdeformed state more symmetric cluster-configurations, and the superdeformed state shows a picture in between. It also turned out that the ^{24}Mg+^{12}C clusterization is allowed in each of these three states, the difference lies in the relative orientations of the deformed clusters.

As for the hyperdeformed state is concerned, in addition to the ^{24}Mg+^{12}C binary configuration the ^{20}Ne+^{16}O turned out to be allowed from among the alpha-like cluster-configurations, which are energetically preferred.

It is remarkable, that these two binary cluster configurations were also selected on the basis of alpha-cluster calculation [4]. The difference in this respect between the two works is that in [5] all the possible binary configurations were studied systematically, and the forbidden or allowed nature of the clusterization was investigated not only from the viewpoint of Harvey's prescription, but also on the basis of the $U(3)$ selection rule. These two methods are complementer to each other in a sense, therefore both of them should be investigated (for a detailed discussion see [10]), otherwise the result might be misleading.

We mention here that a similar study concerning the possible ternary clusterizations of the ground, superdeformed and hyperdeformed states have been performed in [11].

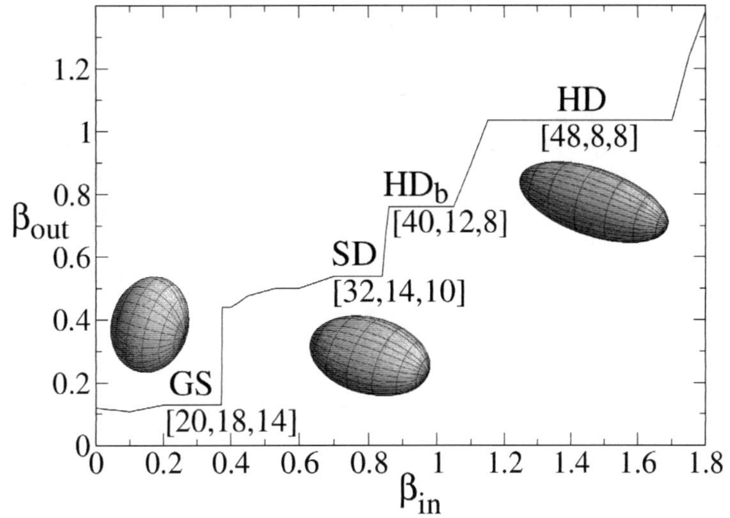

Figure 3. Quadrupole deformation of the ^{36}Ar nucleus from the Nilsson-model with the quasi-dynamical (effective) $U(3)$ quantum numbers at the plateaus. The approximate shape of the GS, SD and HD states are also shown.

3. Experimental verification

In [6] some high-resolution experimental data [12] of the ^{24}Mg+^{12}C elastic scattering has been analyzed. In particular, fifteen angular distribution, measured near the Coulomb-barrier were investigated from the viewpoint of potential scattering, phase-shifts, Regge-poles, and Breit-Wigner terms in the S-matrix. The conclusion was that the fine structure of the cross-section curves can be explained only by the presence of resonances with spins 2-8. These quasi-bound states together with the ^{20}Ne+^{16}O resonances [13] determine a rotational band, as shown in Fig. 2. Its moment of inertia (4.4) is very close to the value, predicted by the alpha-cluster model calculations [4]. It is remarkable that independent experimental studies [14] also showed the importance of these two binary decay channels, although with less detailed analysis of the resonances.

4. HD state from Nilsson-model

After the experimental data seemed to verify the predictions of the cluster studies both on the moment of inertia, and on the reaction channels, it is of great interest to check if the hyperdeformed shape in question shows up also in shell model or mean-field calculations. The first investigation along this line has been carried out within the Nilsson-model in [3]. The stable shapes were obtained from symmetry-considerations, and the result is shown in Fig. 3. This study revealed that, i) in addition to the well-known ground and superdeformed states (which are in line with other results), two candidates are there for the hyperdeformed shape, which are very similar to the simple guess of [5]. One of them (HD_b) has a small triaxiality, while the other one (HD) is cylindrical, and very stable. Furthermore, ii) the HD state obtained in this way has not only similar value for the moment of inertia, than the one predicted from Bloch-Brink-model, and given by the experiment, [4], but also its $U(3)$ symmetry is identical with the previous one.

Journal of Physics: Conference Series **239** (2010) 012006 doi:10.1088/1742-6596/239/1/012006

5. Binary clusterizations

The possible binary cluster-configurations of the shape isomers in ^{36}Ar have been investigated in [3, 5] by taking into account both the requirements of the exclusion principle, and the energy-minimum principle. Some simple clusterizations corresponding to the shape isomers of Fig. 3 are shown in Fig. 1.

As for energetic preference of the allowed binary clusterizations are concerned, it was obtained in [5] from binding energy arguments. The criterium of maximal stability [15], requires the largest value of the summed differences of the measured binding energies and the corresponding liquid drop values: $D(1,2) = [B(1) - B_L(1)] + [B(2) - B_L(2)]$, where $B(i)$ is the experimental binding energy of the ith cluster, while $B_L(i)$ stands for liquid drop value. The alpha-like clusterizations turned out to be more stable than the others.

Here we present the results of a double-folding calculations carried out within the Dinuclear System Model [16]. In this approach effective nucleon-nucleon interactions are applied, which are suitable for structure studies [17]. For the details of this method we refer to the work [18]. The potential energy was determined for the geometrical arrangement obtained from the microscopic considerations. The results are listed in Table 1. For some states different (but close-lying) theoretical candidates are given. In particular e refers to effective symmetry from the Nilsson-calculation, c stands for a cylindrical, while tri for a triaxial shape.

In the energetic preference of the microscopically allowed clusterizations one can observe a general tendency in favour of the lighter clusters: alpha-particle has deeper binding than ^{8}Be, etc. But the relative orientation turned out to be very important, too. For clusterizations, which are present in different shape isomers, like e.g. ^{24}Mg+^{12}C and ^{28}Si+^{8}Be the largest deviation between different geometrical locations is 5.4 MeV (in both cases), which is not smaller than the difference between the averages of the two different cluster-configurations (4.0 MeV).

In [3] the relation of the moments of inertia of the shell model and the corresponding cluster states were discussed. Both of them were calculated classically for rigid bodies, corresponding to the appropriate $U(3)$ symmetries. For the clusterizations a touching configuration was supposed for the (generally deformed) clusters. The conclusion (in agreement with other similar studies) was that this cluster configuration gives larger moment of inertia, than the shell model, and two different ways were discussed for obtaining closer values. With the Dinuclear Molecular System calculation now we are in a position of being able to check a different method. Namely, this consideration provides us with the intercluster distance (R_D) through the minimum of the energy. We have compared these values for the binary clusterizations of Fig. 1 with those of the touching configurations (R_T). It turned out that these values are very close to each other, and consequently so are the corresponding moments of inertia (I_D, and I_T), as well. In particular for the ^{24}Mg+^{12}C clusterization they are R_D (R_T) = 5.2 (4.6), 6.2 (6.6), 6.7 (6.6), 7.9 (7.9) , in units of fm, respectively for the states with increasing deformation. Correspondingly: I_D (I_T) = 3.1-3.6 (2.7-3.2), 4.4-4.7 (4.9-5.0), 5.0-5.2 (4.9-5.0), 6.4-6.5 (6.4-6.5) in units of $10^5 fm^2 MeV/c^2$, and the two values coupled by the sign of "-" appear due to the non-cylindric shape of the configuration. A similar situation was found for the other binary configurations of Fig. 1 as well. For the ^{32}S+^{4}He R_D (R_T) = 6.4 (6.4) and I_D (I_T) = 3.7-4.1 (3.7-4.1). For the ^{20}S+^{16}O R_D (R_T) = 5.8 (5.8) and I_D (I_T) = 3.9 (3.9). For the ^{28}Si(p)+^{8}Be R_D (R_T) = 6.7 (6.8), 8.4 (8.7) and I_D (I_T) = 5.2 (5.3), 7.3 (7.7), respectively, with increasing deformation of the sates. In short, the intercluster distances and moments of inertia obtained from the energy-calculation show a very similar behaviour to those for the touching configuration.

6. Summary and conclusions

In this paper we have reviewed some investigations related to a possible hyperdeformed state of the ^{36}Ar nucleus. It was first predicted from an alpha-cluster model calculation [4], where also a correspondence to the ^{24}Mg+^{12}C and ^{20}Ne+^{16}O resonances was put forward. Then a

Table 1. Orientation and binding energy of binary alpha-like cluster-configurations in the ground and deformed sates of ^{36}Ar. Here Sh stands for shape, while DNS denotes potential energy from the Dinuclear System Model (in MeV). The molecular axis is chosen to be z. s means spherical, p: prolate, o: oblate, t: triaxial, a: symmetry axis. The explanation for the specific orientations are as follows 0*: ^{28}Si symmetry (short) axis in yz plane (closer to z); 1*: ^{24}Mg the largest axis in x direction; 2*: ^{28}Si symmetry (large) axis in xz plane (closer to x); 3*: ^{32}S the largest axis in z direction; 4*: ^{20}Ne symmetry (large) axis in xz plane at 45°; 5*: ^{24}Mg the largest axis in z direction; 6* ^{24}Mg the largest axis in yz plane (closer to y);

State	C_1	Sh	Ratio	Position	C_2	Sh	Ratio	Position	DNS
GS(e) [20, 18, 14]	^{24}Mg	t	1.7 : 1.2 : 1	$x > y > z$	^{12}C	o	1.7 : 1	$a : z$	4.3
	^{28}Si	p	1.6:1	$x = y > z$	^{8}Be	p	2:1	$a : x$	1.0
	^{28}Si	o	1.7:1	$x = y > z$	^{8}Be	p	2:1	$a : x$	−1.7
	^{32}S	t	1.5 : 1.3 : 1	$x = y > z$	^{4}He	s	1	–	−3.7
GS(c) [20, 20, 12]	^{24}Mg	t	1.7 : 1.2 : 1	1*	^{12}C	o	1.7 : 1	$a : z$	1.8
	^{28}Si	p	1.6:1	$x > y > z$	^{8}Be	p	2:1	$a : x$	1.0
	^{28}Si	o	1.7:1	0*	^{8}Be	p	2:1	$a : x$	−1.6
	^{32}S	t	1.5 : 1.3 : 1	$x > y = z$	^{4}He	s	1	–	−3.6
SD(e) [32, 14, 10]	^{24}Mg	t	1.7 : 1.2 : 1	$x = y > z$	^{12}C	o	1.7 : 1	$a : y$	7.2
	^{28}Si	p	1.6:1	$x > z > y$	^{8}Be	p	2:1	$a : z$	3.7
SD(c) [32, 12, 12]	^{24}Mg	t	1.7 : 1.2 : 1	6*	^{12}C	o	1.7 : 1	$a : y$	6.2
	^{28}Si	p	1.6:1	$x = y = z$	^{8}Be	p	2:1	$a : z$	3.3
SD(tri) [32, 16, 8]	^{20}Ne	p	1.6 : 1	$a : x$	^{16}O	s	1	–	8.1
	^{24}Mg	t	1.7 : 1.2 : 1	1*	^{12}C	o	1.7 : 1	$a : x$	6.2
	^{28}Si	p	1.6:1	2*	^{8}Be	p	2:1	$a : z$	2.8
	^{32}S	t	1.5 : 1.3 : 1	3*	^{4}He	s	1	–	−4.3
HD_b [40, 12, 8]	^{20}Ne	p	1.6 : 1	4*	^{16}O	s	1	–	9.4
	^{24}Mg	t	1.7 : 1.2 : 1	5*	^{12}C	o	1.7 : 1	$a : z$	3.4
	^{28}Si	p	1.6:1	$a : z$	^{8}Be	p	2:1	$a : x$	−0.6
HD [48, 8, 8]	^{20}Ne	p	1.6 : 1	$a : z$	^{16}O	s	1	–	8.9
	^{24}Mg	t	1.7 : 1.2 : 1	5*	^{12}C	o	1.7 : 1	$a : x$	6.0
	^{28}Si	p	1.6:1	$a : z$	^{8}Be	p	2:1	$a : z$	2.5

systematic study on the possible binary cluster-configurations [5] showed that these two channels are really the preferred ones for its population. A recent careful analysis of the ^{24}Mg+^{12}C elastic scattering data revealed the existence of low-spin resonances [6], and it turned out, that they form a rotational band together with the ^{20}Ne+^{16}O data (Fig. 2). The moment of inertia is very close to the prediction of the alpha-cluster model. Then from a Nilsson-model calculation a hyperdeformed state was found [3] (Fig. 3), which has a very similar moment of inertia, like that of the experiment and alpha-cluster model. Furthermore, it has an $U(3)$ symmetry, which coincides with that of [4].

All these circumstances make the rotational band of Fig. 2 a very good candidate for the hyperdeformed state of the ^{36}Ar nucleus. For the final conclusion and better understanding we need, of course, more experimental evidence, and theoretical studies. Nevertheless, it seems that Fig. 2 is one of the very few candidates at present for showing evidence for the presence of ground, superdeformed and hyperdeformed shapes in a single nucleus.

Acknowledgments
It is a great honour for the authors to present a paper in a volume which is dedicated to the memory of Professor Marcos Moshinsky. This work was supported by the OTKA (Grant No. K72357), by the MTA-JINR collaboration (Project No. 2009/001), by the NKTH (Hungarian-Spanish collaboration project No. ES-26/2008), and by the MTA-CONACyT exchange programme.

References

[1] Freer M 2007 *Rep. Prog. Phys.* **70** 2149
[2] Svensson C E, Machiavelli A O, Juodagalvis A, Poves A, Ragnarsson I, Aberg S, Appelbe D E, Austin R A E, Baktash C, Ball G C *et al* 2000 *Phys. Rev. Lett.* **85** 2693
[3] Cseh J, Darai J, Sciani W, Otani Y, Lepine-Szily A, Benjamin E A, Chamon L C, Lichtenthaler R 2009 *Phys. Rev.* C **80** 034320
[4] Rae W D M and Merchant A C 1992 *Phys. Lett.* B **279** 207
[5] Cseh J, Algora A, Darai J and Hess P O 2004 *Phys. Rev.* C **70** 034311
[6] Lepine-Szily A, Sciani W, Benjamin E A, Chamon L C, Lichtenthaler R and Otani Y 2008 *J. Phys.: Conf. Series* **111** 012037
Sciani W, Otani Y, Lepine-Szily A, Benjamin E A, Chamon L C, Lichtenthaler R, Darai J and Cseh J 2009 *Phys. Rev.* C **80** 034319
[7] Zhang J, Rae W D M and Merchant A C 1994 *Nucl. Phys.* A **575** 61
[8] Merchant A C *Private communication*
[9] Buck B, Merchant A C and Perez S M (2000) *Few-Body Systems* **29** 53
[10] Cseh J and Darai J 2008 *AIP Conf. Proc.* **1098** 225
[11] Algora A, Cseh J, Darai J and Hess P O 2006 *Phys. Lett.* B **639** 451
[12] Sciani W, Lepine-Szily A, Lichtenthaler R, Fachini P, Gomes L C, Lima G F, Obuti M M, Oliveira J M and Villari A A C 1997 *Nucl. Phys.* A **620** 91
[13] Shimizu J, Yokota W, Nakagawa T, Fukuchi Y, Yamaguchi H, Sato M, Hanashima S, Nagashima Y, Furuno K, Katori K and Kubono S 1982 *Phys. Lett.* B **112** 323
Gai M, Berkowitz G M, Braun-Munzinger P, Jachcinski C M, Ordoñez C E, Renner T R and Uhlhorn C D 1984 *Phys. Rev.* C **30** 925
Freer M and Merchant A C 1997 *J. Phys. G: Nucl. Part. Phys.* **23** 261
[14] Beck C *et al* 2008 *AIP Conf. Proc.* **1098** 207
Beck C *et al* 2009 *Phys. Rev.* C **80** 034604
[15] Buck B, Merchant A C and Perez S M 2000 *Few-Body Systems* **29** 53
Buck B, Merchant A C, Horner M J and Perez S M 2000 *Phys. Rev.* C **61** 024314
[16] Volkov V V 1978 *Phys. Rep.* **44** 93 (1978)
Volkov V V 1982 *Deep inelastic nuclei reactions* (Energoizdat, Moscow)
[17] Migdal A B 1982 *Theory of finite fermi systems and properties of atomic nuclei* (Nauka, Moscow)
[18] Shneidman T M, Adamian G G, Antonenko N V, Jolos R V and Scheid W 2002 *Phys. Lett.* B **526** 322
Shneidman T M, Adamian G G, Antonenko N V, Jolos R V and Scheid W 2003 *Phys. Rev.* C **67** 014313

Single-particle structure of neutron-rich nuclei

J A Cizewski[1], K L Jones[1,2], R L Kozub[3], S D Pain[1,4] and the ORRUBA/RIBENS Collaboration

[1] Department of Physics and Astronomy, Rutgers University, New Brunswick, NJ 08903 USA
[2] Department of Physics and Astronomy, University of Tennessee, Knoxville, TN 37996 USA
[3] Department of Physics, Tennessee Technological University, Cookeville, TN 38505 USA
[4] Physics Division, Oak Ridge National Laboratory, Oak Ridge, TN 37831 USA

E-mail: cizewski@rutgers.edu

Abstract. Neutron transfer (d,p) reactions have been measured with rare isotope beams of ^{132}Sn, ^{130}Sn and ^{134}Te accelerated to ≈4.5 MeV/u interacting with CD_2 targets. Reaction protons were detected in an early implementation of the ORRUBA array of position-sensitive silicon strip detectors. Neutron excitations in the $2f_{7/2}$, $3p_{3/2}$, $3p_{1/2}$ and $2f_{5/2}$ orbitals were populated.

1. Introduction

The properties of nuclei near neutron and proton shell closures provide critical benchmarks in predicting properties of nuclei away from closed shells. Establishing these benchmarks is especially important for neutron-rich nuclei far from stability. The origin of about half of the elements heavier than iron comes from the rapid-neutron capture process of nucleosynthesis, proposed to occur in massive explosions in the cosmos, such as supernova explosions or collisions of neutron stars. Observed r-process abundances cannot be reproduced [1, 2] with traditional nuclear shell and astrophysics models, suggesting a quenching of the shell structure far from stability.

Abundances of heavy elements synthesized in the r process can also be sensitive to neutron capture (n,γ) cross sections [3, 4]. For nuclei with low neutron separation energies, near shell closures or with a large excess of neutrons, neutron capture is dominated by direct capture which depends upon the single-neutron spectroscopic factors of low-spin excitations.

To help to solidify the role of ^{132}Sn with N=82 and Z=50 as a benchmark for extrapolations to even heavier nuclei, single-neutron excitations in ^{133}Sn were measured. Neutron transfer reactions with beams of ^{130,132}Sn and ^{134}Te also provide information needed to calculate direct capture cross sections.

Table 1. Preliminary results from the ^{130,132}Sn and ^{134}Te (d,p) measurements

	Adopted E_x(MeV)	J^π	Preliminary Results E_x(MeV)	ℓ transfer	J^π
^{131}Sn	0.000	$(3/2^+)$	(not observed)		
S_n=5.25 MeV			≈2.7	(3)	$(7/2^-)$
			≈3.4	(1)	$(3/2^-)$
			≈4.0	(1)	$(1/2^-)$
			≈4.7	(3)	$(5/2^-)$
^{133}Sn	0.000	$(7/2^-)$	0.0	(3)	$(7/2^-)$
S_n=2.42 MeV	0.854	$(3/2^-)$	≈0.85	(1)	$(3/2^-)$
	1.561	$(9/2^-)$			
	(1.656)	$(1/2^-)$	≈1.4	(1)	$(1/2^-)$
	2.005	$(5/2^-)$	≈2.0	(3)	$(5/2^-)$
^{135}Te	0.000	$(7/2^-)$	0.0	(3)	$(7/2^-)$
S_n=3.34 MeV	0.659	$(3/2^-)$	≈0.66	(1)	$(3/2^-)$
	1.083	$(1/2^-)$	≈1.00	(1)	$(1/2^-)$
	1.837	$(3/2^-,5/2^-)$	≈1.80		

2. Experimental details

The ^{130,132}Sn and ^{134}Te(d,p) reactions were measured in inverse kinematics at the Holifield Radioactive Ion Beam Facility (HRIBF) at Oak Ridge National Laboratory. Short-lived Sn and Te ions were produced from proton-induced fission of UC and mass separated. To maximize the purity of the Sn beams, sulfur was introduced into the ion source. SnS^+ ions were extracted and subsequently broke up in the charge exchange cell. Negative ^{132}Sn ions (> 90% purity) were accelerated in the 25-MeV tandem Electrostatic Accelerator to energies of 630 MeV. Beam energies for the ^{130}Sn and ^{134}Te studies were 630 and 643 MeV, respectively. The Sn and Te beams interacted with CD_2 targets with effective areal density of $\approx 160\mu g/cm^2$.

Reaction protons were measured in an early implementation of the Oak Ridge Rutgers University Barrel Array (ORRUBA) [5] of position-sensitive silicon strip detectors. At forward angles, detector telescopes were deployed, with ΔE detectors of 65 or $140\mu m$ thickness backed by $1000\mu m$ E detectors. At backward angles $1000\mu m$ E detectors were used. Elastically scattered C ions were stopped in the ΔE detectors. Cross sections of elastically scattered deuterons were used to normalize the data. ORRUBA detectors were supplemented by part of the SIDAR array [6] of silicon strip detectors at back angles. Total angular coverage was 57-92, 90-130, and 143-167 degrees in the laboratory. A micro-channel plate detector was located downstream to provide a timing signal for beam-like recoil particles. To aid in elemental identification, a segmented ion chamber was located further downstream.

3. Results and Discussion

The $2f_{7/2}$, $3p_{3/2}$, $3p_{1/2}$ and $2f_{5/2}$ orbitals are the low-spin configurations above the N=82 gap that are expected to be populated in (d,p) neutron transfer reactions on ^{132}Sn and ^{134}Te. The ^{130}Sn(d,p) reaction could also populate $3s_{1/2}$ or $2d_{3/2}$ excitations below the N=82 gap, as well as configurations above the N=82 gap at higher excitation energies.

Cross sections of reaction protons as a function of angle were measured and compared with distorted-wave Born approximation (DWBA) calculations using the code FRESCO [7]. Stroemich [8] optical potentials were used because they were deduced from extensive elastic scattering measurements on Sn nuclei. These potentials were also used in earlier studies [9] of the ^{124}Sn(d,p) reaction in inverse kinematics. Standard Woods-Saxon potential bound state parameters $R = r_0 A^{1/3}$ with $r_0 = 1.25$ fm and diffuseness $a = 0.65$ were adopted. Spectroscopic factors were deduced by comparing experimental to DWBA-calculated cross sections.

A summary of the preliminary energies and angular momentum transfers from the ^{130,132}Sn and ^{134}Te (d,p) reaction measurements is displayed in Table 1. The preliminary data are taken from [10, 11, 12] and compared to adopted values from [13].

The ^{133}Sn ground ($7/2^-$) and excited states at 854 ($3/2^-$) and 2005 keV ($5/2^-$) had been previously deduced by measuring gamma rays following prompt and beta decay of fission fragments [13]. Evidence for the expected $1/2^-$ state was at best highly tentative, since it was unlikely to have been populated in these studies that preferentially populate higher spin states. The present measurement provides the first clear evidence for a state at ≈1.4 MeV that is a candidate for the $3p_{1/2}$ excitation. Angular distributions to the ground and first excited states are consistent with ℓ=3 and ℓ=1 transfer, respectively. By comparing the magnitudes of measured and DWBA-predicted cross sections, all of the spectroscopic factors for ^{133}Sn are large, close to $S = 1$ [11], as expected for concentrated single-neutron strength in the respective states.

The ground-state region in ^{131}Sn was not significantly populated in the ^{130}Sn(d,p) measurement. However, strong population to states above 2 MeV in excitation was observed, with a pattern of Q-values and relative cross sections very similar to those observed in the ^{132}Sn(d,p) measurement. Preliminary angular distributions for the ≈2.7 and ≈3.4-MeV states agree with $\ell = 3$ and $\ell = 1$ transfer, respectively, to presumably $2f_{7/2}$ and $3p_{3/2}$ states.

There have also been extensive studies of ^{135}Te by prompt and beta decay of fission fragments. Previous studies [13] of excitations in ^{135}Te assigned $f_{5/2}$ character to the first ($5/2^-$) state at 1127 keV. Preliminary analysis of (d,p) angular distribution data supports $7/2^-$, $3/2^-$ and $1/2^-$ spin-parity assignments to the ^{135}Te ground, 659- and 1083-keV states, respectively. The $2f_{5/2}$ strength appears to be fragmented in ^{135}Te, compared to its isotone ^{133}Sn, with little, if any, population of the first $5/2^-$ state at 1127 keV.

Recently Beun and Surman and their colleagues [3, 4] have explored the impact that neutron capture cross sections have on r-process nucleosynthesis abundances. An example of the uncertainty of predictions in this region is ^{130}Sn for which the calculated direct (n,γ) cross section varies by almost 3 orders of magnitude, depending upon the mass model [14]. Beun et al. [3] found that the ^{130}Sn(n,γ) cross section can be especially sensitive to predictions of r-process abundances. By increasing the ^{130}Sn(n,γ) rate by a factor of 10 above standard calculations, r-process abundances of heavy elements can change by as much as 15%. In contrast, r-process abundances are relatively insensitive to the ^{132}Sn(n,γ) cross section [3].

Given sizeable population of candidates for $p_{3/2}$ and $p_{1/2}$ excitations in ^{131}Sn, direct neutron capture on ^{130}Sn could be larger than current estimates and therefore would impact r-process abundance calculations. We are in the process of finalizing confirmation of the $\ell = 1$ transfer to the ≈3.4 and ≈4.0-MeV states in ^{130}Sn and calculating direct neutron capture cross sections.

4. Summary and Future Prospects

Neutron transfer (d,p) reactions have been measured with rare isotope beams of ^{130,132}Sn and ^{134}Te. Neutron $\ell = 1$ and $\ell = 3$ excitations above the N=82 shell gap were populated. These measurements highlight ^{132}Sn as an excellent example of a doubly-closed shell nucleus with relatively pure single-neutron excitations in ^{133}Sn. Understanding $\ell = 1$ excitations also has important implications for predicting r-process abundances.

The prospects for future (d,p) reaction studies with neutron-rich rare isotope beams are bright. Experiments with ^{126,128}Sn beams are planned for HRIBF. A measurement of the ^{130}Sn(^{9}Be,^{8}Beγ) reaction is also planned at HRIBF, to confirm the excitation energies of the ^{131}Sn states above the N=82 gap that were populated in the (d,p) reaction. Later this year, the CARIBU facility [15] at Argonne National Laboratory will enable studies of heavier neutron-rich nuclei from the fission of ^{252}Cf. The Facility for Rare Isotope Beams (FRIB) [16], scheduled to come on line later in this decade, would provide beams of all nuclei along the r-process path up to N$\approx$140.

5. Acknowledgments

The authors would like to thank the members of the ORRUBA/RIBENS collaboration for their invaluable participation in all aspects of this work. The Radioactive Ion Beams for Explosive Nucleosynthesis Studies (RIBENS) is a collaboration of scientists from institutions including Oak Ridge National Laboratory, University of Tennessee-Knoxville, Rutgers University, Tennessee Technological University, Colorado School of Mines, Ohio University, and University of Surrey (UK). Professor F. Nunes provided the FRESCO (d,p) reaction calculations. JAC would also like to acknowledge Professor Marcos Moshinsky for introducing her to nuclear physics and physicists in Mexico.

This work was supported in part by the U.S. Department of Energy under contract numbers DE-FG52-08NA28552 (Rutgers, ORAU), DE-AC05-00OR22725 (ORNL), DE-FG02-96ER40955 (TTU), DE-FG02-96ER40990 (TTU), DE-FG03-93-ER40789 (UTK), DE-FG03-93ER40789 (Mines); the National Science Foundation; and the LDRD program of ORNL.

6. References

[1] Dobaczewski J et al. 1996 *Phys. Rev.* C **53** 2809
[2] Pfeiffer B, Kratz K-L, Thielemann F-K, and Walters W B 2001 *Nucl. Phys.* A **693** 282
[3] Beun J, Blackmon J C, Hix W R, McLaughlin G C, Smith M S and Surman R 2009 *J. Phys. G: Nuclear and Particle Physics* **36** 025201
[4] Surman R, Beun J, McLaughlin G C, and Hix W R 2009 *Phys. Rev.* C **79** 045809
[5] Pain S D et al. 2007 *Nucl. Instrum. Meth. Phys. Res.* B **261** 1122
[6] Bardayan D W et al. 1999 *Phys. Rev. Lett.* **83** 45
[7] Thompson I J 1988 *Comp. Phys. Rep.* **7** 167
[8] Stroemich A et al. 1977 *Phys. Rev.* C **16** 2193
[9] Jones K L et al. 2004 *Phys. Rev.* C **70** 067602
[10] Kozub R L et al. 2008 *Proc. of Science* vol. **NIC-X** 135 and to be published
[11] Jones K L et al. 2010 *Nature* (in press)
[12] Pain S D et al. 2008 *Proc. of Science* vol. **NIC-X** 142 and to be published
[13] http://www.nndc.bnl.gov/ensdf/
[14] Rauscher T et al. 1998 *Phys. Rev.* C **57** 2031
[15] [http://www.phy.anl.gov/atlas/caribu.html]
[16] [http://frib.msu.edu/ and these proceedings]

Journal of Physics: Conference Series **239** (2010) 012008 doi:10.1088/1742-6596/239/1/012008

Proton-neutron pairing correlations in the nuclear shell model

L Yang[1], Y-M Zhao[1], S Pittel[2], B Thakur[2], N Sandulescu[3] and A Poves[4]

[1] Department of Physics, Shanghai Jiao Tong University, Shanghai, 200240, China
[2] Bartol Research Institute and Department of Physics and Astronomy, University of Delaware, Newark, Delaware 19716, USA
[3] Institute of Physics and Nuclear Engineering, 76900 Bucharest, Romania
[4] Universidad Autónoma de Madrid, E-28049, Madrid, Spain

E-mail: pittel@bartol.udel.edu

Abstract. A shell-model study of proton-neutron pairing in $f-p$ shell nuclei using a parametrized hamiltonian that includes deformation and spin-orbit effects as well as isoscalar and isovector pairing is reported. By working in a shell-model framework we are able to assess the role of the various modes of proton-neutron pairing in the presence of nuclear deformation without violating symmetries. Results are presented for ^{44}Ti, ^{46}Ti and ^{48}Cr.

1. Introduction

It is generally believed that *proton − neutron* (*pn*) pairing is important in nuclei with roughly equal numbers of neutrons and protons [1]. The standard technique for treating these correlations is through BCS or HFB approximation, generalized to include the *pn* pairing field in addition to the *nn* and *pp* pairing fields. Questions arise, however, as to whether these methods can adequately represent the physics of the competing modes of pair correlations, without full restoration of symmetries [2].

Important insight into this issue has been achieved recently in the context of exactly-solvable models that include these different pairing modes. Analysis of the SO(8) model, in which isoscalar and isovector pairing act in either a single active orbital or a series of degenerate orbitals, suggests that isospin restoration or equivalently quartet correlations are extremely important, especially near $N = Z$ [2]. More recent studies, carried out for models involving non-degenerate orbitals [3], reinforce earlier conclusions as to where isoscalar pairing correlations should be most important [4, 5]. Furthermore, they make possible the description of deformation, as is critical for systems with $N \approx Z$, by treating the non-degenerate orbitals as Nilsson-like. However, it is still not possible in these models to restore symmetries, either rotational or isospin.

As a consequence, there still remain many open issues concerning the role of the different possible modes of pairing in $N \approx Z$ nuclei. In this work, we report a systematic study of pairing correlations in the context of the nuclear shell model, whereby deformation can be readily included and symmetries maintained. In this way, we are able to address many of the open issues on the role of the various pairing modes in the presence of nuclear deformation.

An outline of the paper is as follows. In Section 2 we describe our model and in Section 3, we report some of the key results we have obtained, which are then summarized in Section 4.

2. Our model

To address in a systematic way the role of pairing correlations in the presence of nuclear deformation, we consider neutrons and protons restricted to the orbitals of the $1f-2p$ shell outside a doubly-magic ^{40}Ca core and interacting via a schematic hamiltonian

$$H = \chi \left(Q \cdot Q + aP^{\dagger} \cdot P + bS^{\dagger} \cdot S + \alpha \sum_i \vec{l}_i \cdot \vec{s}_i \right) \tag{1}$$

where $Q = Q_n + Q_p$ is the mass quadrupole operator, $P^{\dagger}$ creates a correlated $L = 0$, $S = 1$, $J = 1$, $T = 0$ pair and $S^{\dagger}$ creates an $L = 0$, $S = 0$, $J = 0$, $T = 1$ pair. The first term in the hamiltonian produces rotational collective motion, whereas the second and third term are the isoscalar and iosvector pairing interactions, respectively. The last term is the one-body part of the spin-orbit interaction, which splits the $j = l \pm 1/2$ levels with a given l.

We carry out calculations systematically as a function of the various strength parameters. We begin with pure SU(3) rotational motion associated with the $Q \cdot Q$ interaction and then gradually ramp up the various SU(3)-breaking terms to assess how they affect the rotational properties. This includes the isocalar and isovector pairing interactions and the spin-orbit term.

We first consider the nucleus ^{44}Ti, with $N = Z = 2$, and then systematically increase N and Z to study the role of the number of active neutrons and protons, e.g. whether there is an excess of one over the other and whether the nucleus is even-even or odd-mass. The nuclei we have treated are ^{44}Ti ($N = Z = 2$), ^{45}Ti ($N = 2$, $Z = 3$), ^{46}Ti ($N = 2$, $Z = 4$), ^{46}V ($N = 3$, $Z = 3$), and ^{48}Cr ($N = 4$, $Z = 4$). Some of the observables we have studied are (1) the energies and associated BE(2) values of the lowest rotational band, (2) the number of $J^{\pi} = 6^{+}$ $T = 1$ pairs, (3) the number of $J^{\pi} = 0^{+}$, $T = 1$ (S) pairs and (4) the number of $J^{\pi} = 1^{+}$, $T = 0$ (P) pairs. In the following section, we present selected results for ^{44}Ti, ^{46}Ti and ^{48}Cr.

3. Calculations

3.1. Optimal hamiltonian

We first ask whether the hamiltonian (1) has sufficient flexibility to describe the nuclei under investigation. Without making an effort towards an absolute fit, we find that the choice $\chi = -0.5$ MeV, $a = b = 12$, and $\alpha = 20$ gives an acceptable fit to the spectra of all the nuclei we have considered. This is illustrated in figure 1 for ^{44}Ti, ^{46}Ti and ^{48}Cr. As can be seen, the well-known non-rotational character of ^{44}Ti is reproduced by our calculations, as are the highly rotational patterns seen experimentally for the heavier nuclei. As we will see later, even the experimentally observed backbend in ^{48}Cr is acceptably reproduced with this hamiltonian. We refer to the choice $a = b$ in the *optimal* hamiltonian as the SU(4) choice, from the dynamical symmetry that derives from this choice of parameters in the SO(8) model.

3.2. ^{44}Ti

We next turn to the nucleus ^{44}Ti, with two active neutrons and two active protons. In figure 2, we show the calculated energy splittings $E_I - E_{I-2}$ associated with the ground-state band as a function of the strength parameters a and b that define the isoscalar and isovector pairing interactions, respectively. For these calculations we assumed a quadrupole strength of $\chi = -0.05$ MeV and no spin-orbit interaction. What we see is that the isoscalar and isovector pairing interactions have precisely the same effect on the properties of the ground state rotational band, *in the absence of any spin-orbit interaction.* Precisely the same conclusion derives when we consider the effect of isoscalar and isovector pairing on other observable properties.

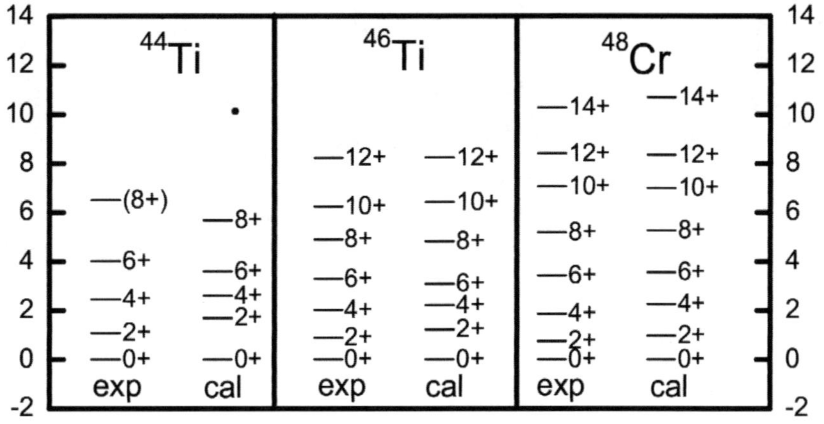

Figure 1. Comparison of experimental spectra for ^{44}Ti, ^{46}Ti and ^{48}Cr with the calculated spectra obtained using the *optimal* hamiltonian described in the text. All energies are in MeV.

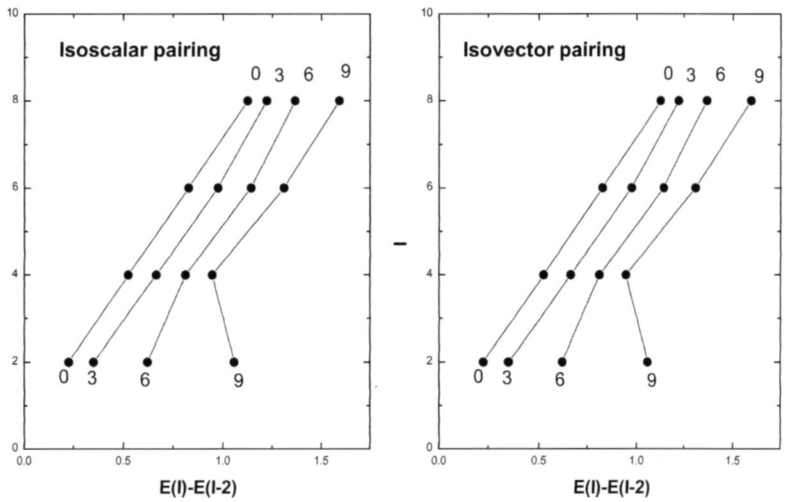

Figure 2. Spectra of the ground state band of ^{44}Ti as a function of the strength of the isoscalar pairing interaction (left panel) and of the isovector pairing interaction (right panel), in each case with no spin-orbit term present.

Some other results from our study of ^{44}Ti are illustrated in figure 3. In panel (a), we show the spectrum that derives solely from turning on a strong spin-orbit force. We see that the spectrum is still highly rotational, despite the fact that the resulting single-particle energies are no longer SU(3)-like. To obtain the physical spectrum with a non-rotational character, it is thus essential to have pairing. It is usually accepted that it is the non-SU(3) order of the single-particle levels that is responsible for the non-rotational character seen [6], a conclusion not supported by our results. In panel (b), we address the issue of how isoscalar pairing is affected by inclusion of a spin-orbit force. We see that the number of isoscalar pairs – defined as the expectation value of $P^{\dagger} \cdot P$ – decreases rapidly with an increasing spin-orbit interaction,

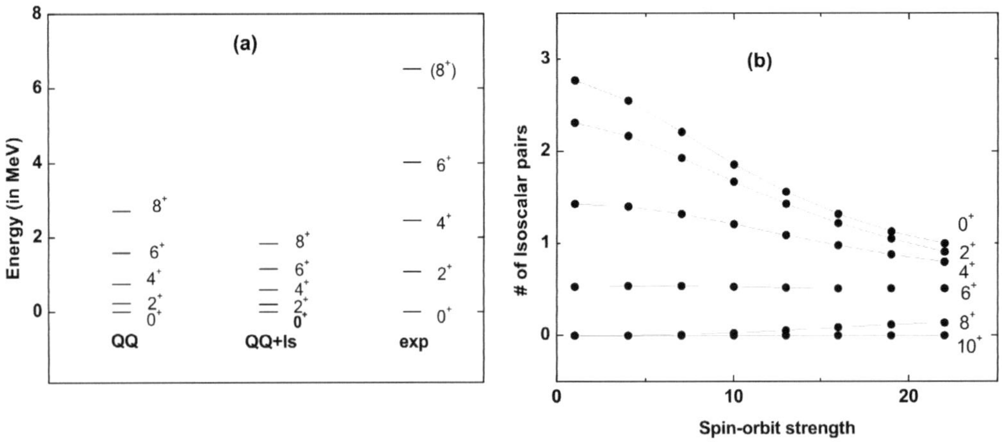

Figure 3. Results from calculations carried out for ^{44}Ti. (a) Comparison of he experimental spectrum with spectra obtained with a pure $Q \cdot Q$ interaction and with both a $Q \cdot Q$ interaction and a spin-orbit term. (b) Number of isoscalar pairs as a function of the spin-orbit strength, for the various states in the ground band.

especially for the lowest angular momentum states of the ground band. The mechanism whereby the spin-orbit interaction suppresses isoscalar pairing has been discussed recently in ref. [7].

3.3. ^{46}Ti

Next we turn to ^{46}Ti with two additional neutrons present. Here too we compare the effect of the isoscalar and isovector pairing interactions on deformation, showing the results in figure 4 with no spin-orbit term present. Here the effect of isoscalar pairing is strongly suppressed relative to isovector pairing, suggesting that even without a spin-orbit term isoscalar pairing is very strongly focused on those nuclei with $N = Z$ with a slight excess being sufficient to suppress this pairing mode.

3.4. ^{48}Cr

Lastly, we turn to ^{48}Cr, which again has $N = Z$, but now with two quartet-like structures present. Here we assume as our starting point both the optimal quadrupole-quadrupole force and one-body spin-orbit force and then ramp up the two pairing strengths from zero to their optimal values. The results are illustrated in figure 5, for scenarios in which we separately include isoscalar pairing, isovector pairing and and SU(4) pairing with equal strengths.

As a reminder, the experimental spectrum for ^{48}Cr shows a backbend near $I = 12$, which as noted earlier is reproduced by our *optimal* hamiltonian. The results of figure 5 make clear that (a) the backbend cannot be reproduced with pure isoscalar pairing, but requires isovector pairing as well and (b) there is no significant difference between the results obtained with pure isovector pairing and SU(4) pairing. The backbend in ^{48}Cr was discussed extensively in the context of a shell-model study with a fully realistic hamiltonian in [8], where it was first shown to derive from isovector pairing. Our results are in agreement with that earlier conclusion.

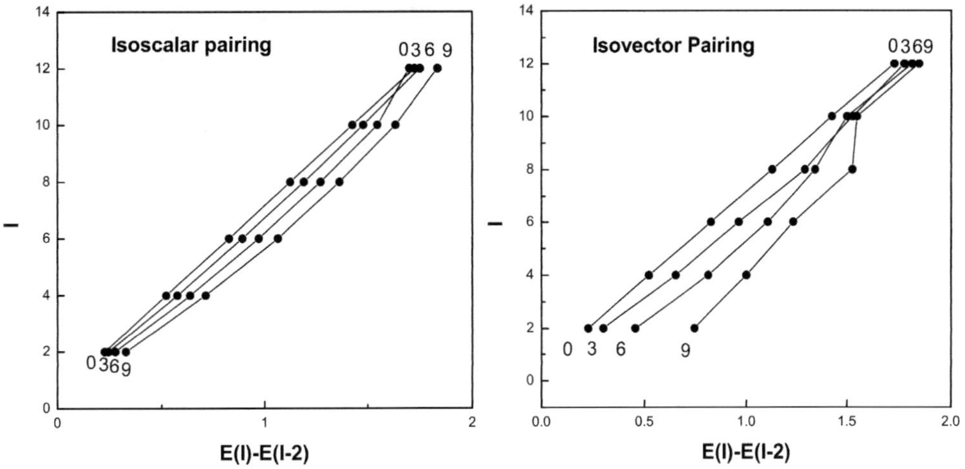

Figure 4. Spectra of the ground state band of ^{46}Ti as a function of the strength of the isoscalar pairing interaction (left panel) and of the isovector pairing interaction (right panel), in each case with no spin-orbit term present.

4. Summary and Concluding Remarks

In this work, we have reported a shell-model study of proton-neutron pairing in $f-p$ shell nuclei using a parametrized hamiltonian that includes deformation and spin-orbit effects as well as both isoscalar and isovector pairing. By working in a shell-model framework we are able to assess the role of the various modes of proton-neutron pairing in the presence of nuclear deformation without violating symmetries.

We first showed that our parametrized hamiltonian has enough flexibility to be able to provide a reasonable description of the evolution of nuclear structure properties in this region. We then probed the role of the various modes of pairing on deformation with or without a spin-orbit term. We did this as a function of the number of neutrons and protons, so as to assess the role both of a neutron excess and of the number of active particles.

Some of the conclusions that emerged are: (1) in the absence of a spin-orbit term, isoscalar and isovector pairing have identical effects at $N = Z = 2$, but that isoscalar pairing ceases to have an appreciable effect for nuclei with just two excess neutrons; (2) the non-rotational character of ^{44}Ti cannot be explained solely in terms of spin-orbit effects but requires pairing for its understanding; (3) in the presence of a spin-orbit interaction, isoscalar pairing is suppressed even at $N = Z$, and (4) the known backbend in ^{48}Cr has its origin in isovector pairing.

Acknowledgments

This paper is dedicated by one of the authors (S.P.) to the memory of Marcos Moshinsky, colleague, teacher and friend. The work reported herein began while S.P. and N.S. were visiting the Consejo Superior de Investigaciones Científicas in Madrid, whose hospitality is gratefully acknowledged. Much of it was carried out while L.Y. was visiting the Bartol Research Institute of the University of Delaware, whose hospitality is likewise acknowledged. The work of S.P., B.T. and L.Y. was supported in part by the National Science Foundation under grant #s PHY-0553127 and PHY-0854873 and the work of N.S. by the Romanian Ministry of Education and

Journal of Physics: Conference Series **239** (2010) 012008 doi:10.1088/1742-6596/239/1/012008

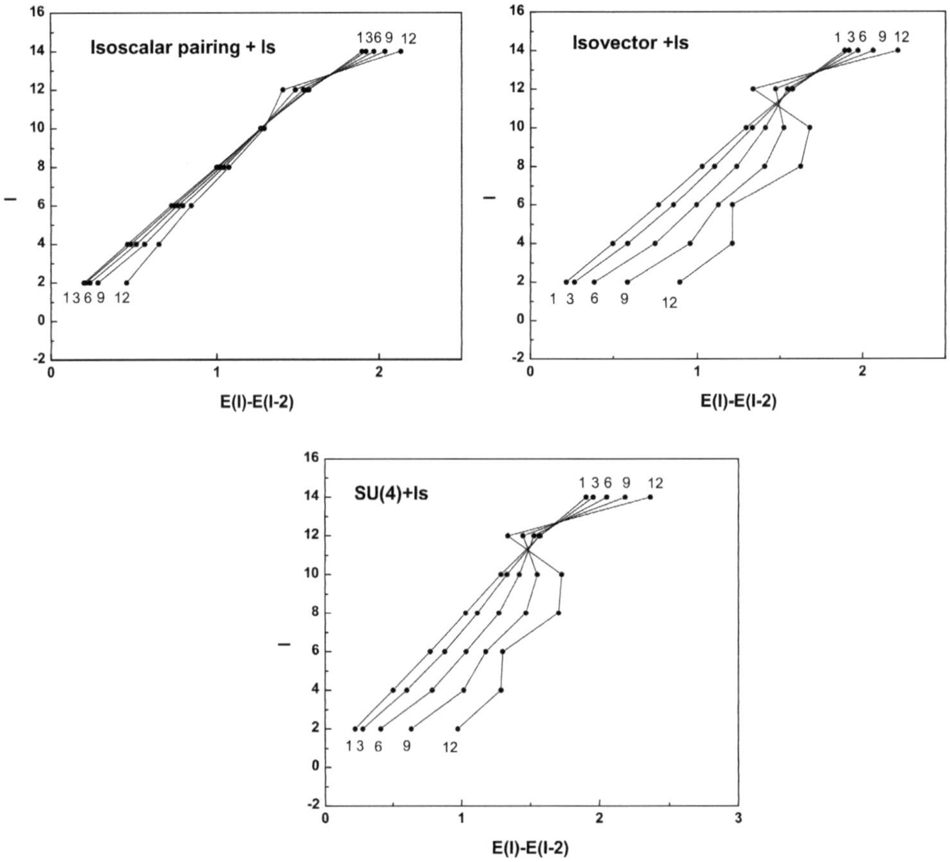

Figure 5. Calculated splittings in ^{48}Cr ground band, for isovector, isoscalar, and SU(4) pairing, respectively, as described in the text.

Research through CNCSIS grant Idei nr. 270.

References

[1] Goodman A L 1979 *Adv. Nucl. Phys.* **11** (1979) 263
[2] Dobeš J and Pittel S 1998 *Phys. Rev.* C **57** 688
[3] Errea B 2009 Ph.D. thesis, Consejo Superior de Investigaciones Científicas, unpublished
[4] Lerma S, Errea B, Dukelsky J and Satula W 2007 *Phys. Rev. Lett.* **99** 064318
[5] Sandulescu N, Errea B, Dukelsky J 2009 *Phys. Rev.* C **80** 044335
[6] Bhatt K H and McGrory J B 1971 *Phys. Rev.* C **3** 2293
[7] Bertsch G F and Baroni S 2009 Suppression of isoscalar pairing (*Preprint* nucl-th/0904.2017v2)
[8] Poves A and Martinez-Pinedo G 1998 *Phys. Lett.* B **430** 203

XXXIII Symposium on Nuclear Physics IOP Publishing
Journal of Physics: Conference Series **239** (2010) 012009 doi:10.1088/1742-6596/239/1/012009

Structure of the nucleon in the unquenched quark model

Roelof Bijker[1] **and Elena Santopinto**[2]

[1] Instituto de Ciencias Nuclares, Universidad Nacional Autónoma de México, AP 70-543, 04510 Mexico DF, Mexico

[2] I.N.F.N., Sezione di Genova, via Dodecaneso 33, 16164 Genova, Italy

E-mail: bijker@nucleares.unam.mx

Abstract. We discuss the flavor asymmetry of the nucleon sea and the spin content of the proton in an unquenched quark model. It is shown that the inclusion of hadron loops leads automatically to an excess of $\bar{d}$ over $\bar{u}$ and introduces a sizeable contribution of orbital angular momentum to the spin of the proton and the Λ hyperon. Special attention is paid to the symmetries of the unquenched quark model.

1. Introduction

One of the challenges of hadronic physics is to understand the structure of the nucleon and its excited states in terms of effective degrees of freedom and, at a more fundamental level, the emergence of these effective degrees of freedom from QCD, the underlying theory of quarks and gluons [1]. The constituent quark model (CQM) is an effective model that has been very successful in explaining hadron properties in terms of system of constituent quarks and antiquarks, qqq for baryons and $q\bar{q}$ for mesons. Nevertheless, there is compelling evidence for the existence of exotic degrees of freedom (other than valence quarks) in hadrons, in particular for the need to include the effects of quark-antiquark pair creation. The importance of quark-antiquark configurations (or higher Fock components in baryon wave functions) is evident from measurements of the $\bar{d}/\bar{u}$ asymmetry in the nucleon sea [2], parity-violating electron scattering experiments [3, 4], the proton spin crisis [5], as well as analysis of helicity amplitudes [6] and strong couplings of baryon resonances [7, 8].

The aim of this contribution is to study the flavor asymmetry and the spin of the proton in an unquenched quark model in which the effects of quark-antiquark pair creation ($u\bar{u}$, $d\bar{d}$ and $s\bar{s}$) are taken into account in an explicit form via a $^3P_0$ coupling mechanism [9]. In addition, we show some predictions for the other octet baryons.

2. Unquenched quark model

The present work on the unquenched quark model for baryons is motivated by earlier studies on extensions of the quark model that employ a $^3P_0$ model for the $q\bar{q}$ pair creation [10, 11]. Our approach is based on a CQM to which the quark-antiquark pairs with vacuum quantum numbers are added as a perturbation [9, 11]. The pair-creation mechanism is inserted at the quark level and the one-loop diagrams are calculated by summing over all possible intermediate states. Under these assumptions, the baryon wave function consists of a zeroth order three-quark configuration $\mid A\rangle$ plus a sum over all possible higher Fock components due to the creation of $^3P_0$ quark-antiquark pairs

$$\mid \psi_A\rangle = \mathcal{N}\left[\mid A\rangle + \sum_{BClJ} \int d\vec{k} \mid BC\vec{k}\,lJ\rangle \frac{\langle BC\vec{k}\,lJ \mid T^\dagger \mid A\rangle}{M_A - E_B - E_C}\right] , \quad (1)$$

where B and C represent the intermediate baryon and meson, respectively. The operator $T^\dagger$ creates a $^3P_0$ quark-antiquark pair in a flavor singlet state [9, 12]. Therefore, the $SU(3)$ flavor symmetry of the valence quark configuration $\mid A\rangle$ is broken by the quark-antiquark pairs via the energy denominator, but the $SU(2)$ isospin symmetry is still preserved. In the closure limit, the energy denominator reduces to a constant and the higher Fock components of the baryon wave function have the same flavor symmetry as the valence quark configuration.

In order to calculate the effects of quark-antiquark pairs on an observable, one has to evaluate the contribution of all possible intermediate states. We developed a combination of group theoretical and computational techniques which makes it possible to perform the sum over intermediate states up to saturation and not only for the first few shells as in previous studies [10, 11]. In addition, it allows the evaluation of the contributions of quark-antiquark pairs for any initial baryon (ground state or resonance) and for any flavor of the $q\bar{q}$ pair (not only $s\bar{s}$ as in [11], but also $u\bar{u}$ and $d\bar{d}$), and for any model of baryons and mesons, as long as their wave functions are expressed in the basis of harmonic oscillator wave functions [9].

The results of the unquenched calculations presented in this contribution are obtained by using a harmonic oscillator quark model for both baryons and mesons. The sum over intermediate states is carried out over five oscillator shells for both the intermediate baryons and mesons. In the unquenched calculations all parameters are taken from the literature without attempting to optimize their values in order to improve the agreement with experimental data [9]. Obviously, the unquenching of the quark model has to be done in such a way as to maintain the phenomenological successes of the CQM. In applications to mesons, it was shown that the inclusion of quark-antiquark pairs does not destroy the good CQM results [13] and preserves the OZI hierarchy [14]. In a similar fashion, we showed that the CQM results for the magnetic moments of the octet baryons also hold in the unquenched quark model [9].

3. Flavor asymmetry

The flavor content of the nucleon sea provides an important test for models of nucleon structure. A flavor symmetric sea leads to the Gottfried sum rule $S_G = 1/3$, whereas any deviation from this value is an indication of the $\bar{d}/\bar{u}$ asymmetry of the nucleon sea, thus providing evidence of the existence of higher Fock components (such as $qqq - q\bar{q}$ configurations) in the proton wave function. The first clear evidence of a violation of the Gottfried sum rule came from the New Muon Collaboration (NMC) [15], which was later confirmed by Drell-Yan experiments [16, 17] and a measurement of semi-inclusive deep-inelastic scattering [18]. All experiments show evidence that there are more $\bar{d}$ quarks in the proton than there are $\bar{u}$ quarks [2].

The flavor asymmetry is related to the Gottfried integral for the difference of the proton and neutron electromagnetic structure functions

$$S_G = \int_0^1 \frac{F_2^p(x) - F_2^n(x)}{x} dx = \frac{1}{3} - \frac{2}{3}\int_0^1 \left[\bar{d}_p(x) - \bar{u}_p(x)\right] dx . \quad (2)$$

Under the assumption of a flavor symmetric sea one obtains the Gottfried sum rule $S_G = 1/3$. The final NMC value is 0.2281 ± 0.0065 at $Q^2 = 4$ $(\mathrm{GeV/c})^2$ for the Gottfried integral over the range $0.004 \leq x \leq 0.8$ [15], which implies a flavor asymmetric sea. The observed flavor asymmetry is far too large to be accounted for by processes that can be described by QCD in

Journal of Physics: Conference Series **239** (2010) 012009 doi:10.1088/1742-6596/239/1/012009

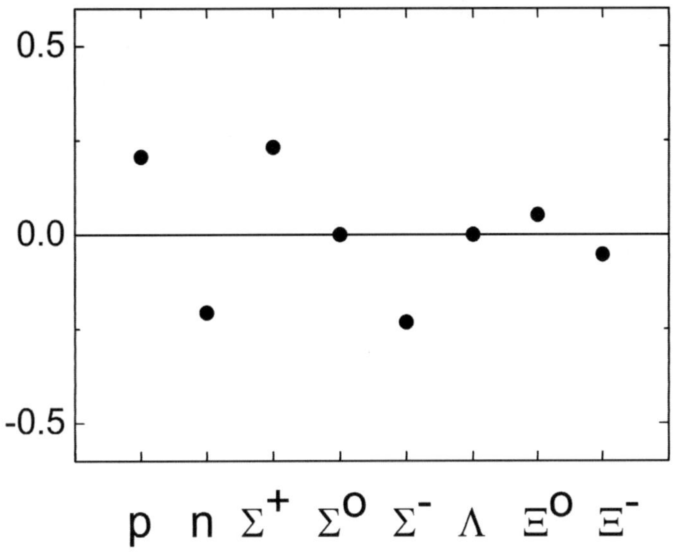

Figure 1. Flavor asymmetry of octet baryons

perturbative regime and therefore has to attributed to non-perturbative QCD mechanisms. It was shown in the framework of the meson cloud model, that the coupling of the nucleon to the pion cloud provides a mechanism that is able to produce a flavor asymmetry due to the dominance of $n\pi^+$ among the virtual configurations [19].

In the unquenched quark model, the flavor asymmetry of the proton can be calculated from the difference of the number of $\bar{d}$ and $\bar{u}$ sea quarks in the proton

$$\mathcal{A}(p) = N_{\bar{d}}(p) - N_{\bar{u}}(p) = \int_0^1 \left[\bar{d}_p(x) - \bar{u}_p(x)\right] dx \ . \tag{3}$$

Even in absence of explicit information on the (anti)quark distribution functions, the integrated value can be obtained directly from the left-hand side of Eq. (3). The corresponding value for the Gottfried integral of Eq. (2) is 0.21, in qualitative agreement with the experimental results. The flavor asymmetries of the octet baryons are related by isospin symmetry. Since the operator $\hat{N}_{\bar{d}} - \hat{N}_{\bar{u}}$ is an isovector in isospin space, the flavor asymmetries satisfy [20]

$$\mathcal{A}(A) = \langle \psi_A(I, I_3) \mid \hat{N}_{\bar{d}} - \hat{N}_{\bar{u}} \mid \psi_A(I, I_3) \rangle = \frac{I_3}{I} \langle \psi_A(I, I) \mid \hat{N}_{\bar{d}} - \hat{N}_{\bar{u}} \mid \psi_A(I, I) \rangle \ , \tag{4}$$

where I and I_3 denote the isospin and its projection of baryon A. As a consequence, the excess of $\bar{d}$ over $\bar{u}$ in the proton is related to the excess of $\bar{u}$ over $\bar{d}$ in the neutron, $\mathcal{A}(p) = -\mathcal{A}(n)$. Similar relations hold for the other octet baryons: $\mathcal{A}(\Sigma^+) = -\mathcal{A}(\Sigma^-)$, $\mathcal{A}(\Xi^0) = -\mathcal{A}(\Xi^-)$ and $\mathcal{A}(\Lambda) = \mathcal{A}(\Sigma^0) = 0$. These relations help to explain the regularities observed for the flavor asymmetry as shown in Figure 1.

Finally, in the unquenched quark model the flavor asymmetry of the proton is predicted to be of the same order as that of the Σ hyperon and much larger than that of the cascade particle, $\mathcal{A}(p) \sim \mathcal{A}(\Sigma^+) \gg \mathcal{A}(\Xi^0)$. This behavior is very different from that obtained in other models, such as the cloudy bag model, the chiral quark model and the balance model which all predict that the asymmetry of the cascade particle is equal or larger than that of the proton $\mathcal{A}(\Xi^0) \geq \mathcal{A}(p)$ [21]. In order to distinguish between the predictions of the different models and to obtain a better understanding of non-perturbative structure of QCD new experiments are needed to measure the flavor asymmetry of hyperons. In particular, the flavor asymmetry of charged Σ hyperons can obtained from Drell-Yan experiments using charged hyperon beams on the proton [22].

4. Proton and Lambda spin

The observation by European Muon Collaboration that the total quark spin constitutes a rather small fraction of the spin of the nucleon [23] sparked an enormous interest in the spin structure of the proton [5]. Recent experiments show that approximately one third of the proton spin is carried by the quarks and antiquarks [24, 25], and that the gluon contribution is rather small (either positive or negative) and compatible with zero [26]. This rules out the possibility that most of the missing spin be carried by the gluon and indicates that the origen of the missing spin of the proton has to be attributed to other mechanisms.

In the unquenched quark model, the effect of hadron loops on the fraction of the proton spin carried by the quark (antiquark) spins and orbital angular momentum can be studied in an explicit way [27]. As in other effective models [5], gluonic effects associated with the axial anomaly are not included, and therefore the contribution from the gluons is missing from the outset. The total spin of the proton can then be written as the sum of the contributions from the quark (and antiquark) spins and orbital angular momentum

$$\frac{1}{2} = \frac{1}{2}\Delta\Sigma + \Delta L = \frac{1}{2}\left(\Delta u + \Delta d + \Delta s\right) + \Delta L \ . \tag{5}$$

Table 1 shows that the inclusion of the quark-antiquark pairs has a dramatic effect on the spin content of the proton. Whereas in the CQM the proton spin is carried entirely by the (valence) quarks, in the unquenched calculation the contributions of the valence quark spins, the sea quark spins and the orbital angular momentum to the proton spin are comparable in size and equal to approximately 38, 30 and 32 %, respectively. While the orbital angular momentum arises almost entirely from the $N\pi$ and $\Delta\pi$ channels, the contribution of the sea quark spins is dominated by the intermediate vector mesons, for which the convergence of the sum over intermediate states is much slower. Therefore, the sum was carried out over five oscillator shells for both the intermediate baryons and mesons [9].

The results of the HERMES analysis are presented in the column labeled DIS [24] and those of the naive quark model in the column labeled CQM. The importance of orbital angular momentum to the proton spin was discussed many years ago by Sehgal [28] and Ratcliff [29] in the context of the quark-parton model and, more recently, by Myhrer and Thomas in framework of the bag model [30].

The experimental data on the spin structure of the proton have raised many questions about the contributions of valence and sea quarks, gluons and orbital angular momentum to the proton spin. In this respect it is of interest to investigate the spin structure of other octet baryons, in particular that of the Λ hyperon. In the naive CQM, the Λ spin is carried entirely by the strange quark, which makes it a clean example to study the spin structure of baryons. Moreover, its polarization can be measured from the nonleptonic decay $\Lambda \rightarrow p\pi$ [31]. In summary, a study of the spin structure of the Λ hyperon is not only interesting in its own right, but also may shed light on the origen of the spin crisis of the proton.

In most studies, additional assumptions had to be made about the sea quarks in order to get an estimate of the spin content of the Λ hyperon. For example, the assumption that both valence and sea quarks are related by $SU(3)$ flavor symmetry, allows to express the spin content

Table 1. Contribution of Δu, Δd , Δs, $\Delta\Sigma = \Delta u + \Delta d + \Delta s$ and ΔL to the spin of the proton and the Λ hyperon in the unquenched constituent quark model (UCQM).

		CQM	DIS	UCQM		
				Valence	Sea	Total
p	Δu	4/3	0.842	0.504	0.594	1.098
	Δd	−1/3	−0.427	−0.126	−0.291	−0.417
	Δs	0	−0.085	0.000	−0.005	−0.005
	$\Delta\Sigma$	1	0.330	0.378	0.298	0.676
	$2\Delta L$	0		0.000	0.324	0.324
	$2J$	1		0.378	0.622	1.000
Λ	Δu	0	−0.159	0.000	−0.055	−0.055
	Δd	0	−0.159	0.000	−0.055	−0.055
	Δs	1	0.647	0.422	0.539	0.961
	$\Delta\Sigma$	1	0.330	0.422	0.429	0.851
	$2\Delta L$	0		0.000	0.149	0.149
	$2J$	1		0.422	0.578	1.000

of the Λ hyperon in terms of that of the proton [31]

$$\begin{aligned} (\Delta u)_\Lambda = (\Delta d)_\Lambda &= (\Delta u + 4\Delta d + \Delta s)_p /6 \;, \\ (\Delta s)_\Lambda &= (2\Delta u - \Delta d + 2\Delta s)_p /3 \;. \end{aligned} \tag{6}$$

An analysis of the experimental DIS data for the proton [24, 25] in combination with Eq. (6) shows that the strange quarks (and antiquarks) carry about 65 % of the Λ spin, while the up and down quarks (and antiquarks) account for a negative polarization of –32 %.

In the unquenched quark model there is no need to make additional assumptions about the nature of the sea. Since the $SU(3)$ flavor symmetry is broken by the sea quarks, Eq. (6) does not hold for the unquenched calculations. Table 1 shows that the unquenched quark model gives rise to a negatively polarized sea of up and down quarks and a large positively polarized sea of strange quarks. Other theoretical studies, such as the chiral quark-soliton model [32], lattice QCD [33] and QCD sum rules [34], also give a negative polarization of the up and down quarks, Finally, the contribution of quark spins for the Λ is found to be larger than that for the proton, $(\Delta\Sigma)_\Lambda > (\Delta\Sigma)_p$, which is a result of $SU(3)$ flavor breaking by the sea quarks.

5. Summary and conclusions

In this contribution, we discussed the effects of hadron loops an unquenched quark model for baryons in which the quark loops are taken into account via a $^3P_0$ pair creation model. The ensuing unquenched quark model is valid for any baryon (or baryon resonance), includes all light flavors of the quark-antiquark pairs ($u\bar{u}$, $d\bar{d}$ and $s\bar{s}$), and can be used in combination with any CQM, as long as its wave functions are expressed in a harmonic oscillator basis.

The inclusion of the $q\bar{q}$ pairs leads automatically to an excess of $\bar{d}$ over $\bar{u}$ quarks, in agreement with the observed flavor asymmetry of the proton. In addition, it leads to a sizeable contribution of the orbital angular momentum to the spin of the proton ($\sim$ 32 %) and the Λ hyperon ($\sim$ 15 %). In addition, there is a large contribution from the spins of the sea quarks due to intermediate vector mesons for the proton ($\sim$ 30 %) and the Λ ($\sim$ 43 %). We note, that this contribution is absent in meson-cloud models. In order to be able to distinguish between the predictions of different models of hadron structure and to obtain a better understanding of non-perturbative structure of QCD new experiments are needed to measure the flavor asymmetry and spin content of other octet baryons.

In conclusion, the formulation of an unquenched quark model in which the effects of quark-antiquark pairs are included explicitly in a general and consistent way, may provide an important step forward in CQM studies and increase considerably its range of applicability. Future work includes applications to radii, strong couplings, electromagnetic couplings and transition form factors of baryon resonances.

Acknowledgments

This work was supported in part by a grants from INFN, Italy and CONACyT, Mexico.

References

[1] Isgur N 1997 *Nucl. Phys.* A **623** 37c
[2] Garvey G T and Peng J C 2001 *Prog. Part. Nucl. Phys.* **47** 203
[3] Acha A *et al* 2007 *Phys. Rev. Lett.* **98** 032301
[4] Baunack S *et al* 2009 *Phys. Rev. Lett.* **102** 151803
[5] Bass S D 2008 *The spin structure of the proton* (Singapore: World Scientific)
[6] Aznauryan I G *et al* 2008 *Phys. Rev.* C **78** 045209
Aznauryan I G *et al* 2009 *Phys. Rev.* C **80** 055203
[7] Capstick S and Roberts W 1994 *Phys. Rev.* D **49** 4570
[8] Bijker R, Iachello F and Leviatan A 1997 *Phys. Rev.* D **55** 2862
Bijker R, Iachello F and Leviatan A 2000 *Ann. Phys. (N.Y.)* **284** 89
[9] Bijker R and Santopinto E 2009 *Phys. Rev.* C **80** 065210
[10] Zenczykowski P 1986 *Ann. Phys. (N.Y.)* **169** 453
[11] Geiger P and Isgur N 1997 *Phys. Rev.* D **55** 299
[12] Roberts W and Silvestre-Brac B 1992 *Few-Body Systems* **11** 171
[13] Geiger P and Isgur N 1990 *Phys. Rev.* D **41** 1595
[14] Geiger P and Isgur N 1993 *Phys. Rev.* D **47** 5050
[15] Amaudruz P *et al* 1991 *Phys. Rev. Lett.* **66** 2712
Arneodo M *et al* 1997 *Nucl. Phys.* B **487** 3
[16] Baldit A *et al* 1994 *Phys. Lett.* B **332** 244
[17] Towell R S *et al* 2001 *Phys. Rev.* D **64** 052002
[18] Ackerstaff K *et al* 1998 *Phys. Rev. Lett.* **81** 5519
[19] Thomas A W 1983 *Phys. Lett.* B **126** 97
Henley E M and Miller G A 1990 *Phys. Lett.* B **251** 453
[20] Bijker R and Santopinto E 2008 *Proceedings of the 11th Workshop on the Physics of Excited Nucleons, NSTAR 2007* ed H W Hammer *et al* (Berlin: Springer) p 166 (*Preprint* arXiv:0809.2299)
[21] Shao L, Zhang Y J and Ma B Q 2010 *Phys. Lett.* B **686** 136
[22] Alberg M, Henley E M, Ji X and Thomas A W 1996 *Phys. Lett.* B **389** 367
Alberg M, Falter T and Henley E M 1998 *Nucl. Phys.* A **644** 93
[23] Ashman J *et al* 1988 *Phys. Lett.* B **206** 364
[24] Airapetian A *et al* 2007 *Phys. Rev.* D **75** 012007
[25] Alexakhin V Yu *et al* 2007 *Phys. Lett.* B **647** 8
[26] Bradamante F 2008 *Prog. Part. Nucl. Phys.* **61** 229
[27] Bijker R and Santopinto E 2007 *AIP Conference Proceedings* **947** 168
[28] Sehgal L M 1974 *Phys. Rev.* D **10** 1663
[29] Ratcliffe P G 1987 *Phys. Lett.* B **192** 180
[30] Myhrer F and Thomas A W 2008 *Phys. Lett.* B **663** 302
[31] Jaffe R L 1996 *Phys. Rev.* D **54** R6581
[32] Kim H C, Praszalowicz M and Goeke K 2001 *Acta Phys. Pol.* B **32** 1343
[33] Göckeler M, Horsley R, Pleiter D, Rakow P E L, Schaefer S, Schäfer A and Schierholz G 2002 *Phys. Lett.* B **545** 112
[34] Erkol G and Oka M 2009 *Phys. Rev.* D **79** 114028

Neutral current neutrino-nucleon interaction and the $\Delta(1232)$

A Mariano[1], C Barbero[1] and G López Castro[2]

[1] Departamento de Física, Facultad de Ciencias Exactas, Universidad Nacional de La Plata, cc. 67, 1900 La Plata, Argentina

[2] Departamento Física, Centro de Investigación y de Estudios Avanzados del IPN, Apdo. Postal 14-740, 07000 México, D.F., México

E-mail: mariano@fisica.unlp.edu.ar

Abstract. Neutrino-induced one-pion production off nucleons and nuclei in the intermediate energy region, is a source of relevant data on hadronic structure. On the other hand, a proper understanding of these processes is very important in the analysis of neutrino oscillation experiments, where the weak production of pions is the main background. Then, it is important develop a consistent model to have under control the charged (CC) and neutral current (NC) pion production cross sections. Working within an isobar model previously developed to describe elastic and radiative pion-nucleon cross sections and pion photoproduction reactions, using the axial form factors (FF) already obtained from the study of the CC neutrino-nucleon currents, we analyze its predictive power in the NC current production case.

1. Introduction

The interest on inelastic weak pion production processes from $\nu N \to l\pi N'$ reactions is twofold. They are a powerful tool to study the hadron structure with high intensity neutrino beams and make possible to investigate the axial structure of the nucleon and baryon resonances. This information will enlarge our view of hadron structure beyond what is presently known about electromagnetic form factors (FF) from JLAB. For example FERMILAB has the NUMI beam and MINERνA(Main INjector ExpeRiment for ν-A) is a neutrino detector designed to study ν-nucleus interactions with unpreceding detail [1]. It will probe the four-momentum transfer squared (Q^2) dependence of the nucleon axial FF and also study neutrino induced πN states dominated by resonances excitation. Pions are mainly produced by them and these reactions can be used to extract information on nucleon-to-resonance axial transition form factors through hadron effective models. Then, these determinations could be compared with those obtained from static quark models or lattice QCD calculations. In addition MINERνA has several nuclear targets, allowing a study of nuclear effects in neutrino interactions. Another fundamental question is the strangeness content of the nucleon spin which can be best unraveled with NC quasielastic ν scattering (NCQE) $\nu N \to \nu N$ (FINeSSE proposal) [2].

On the other hand they have become a very important element in the analysis of neutrino oscillation experiments which searching for the distorsion in the neutrino flux at a detector positioned far away from the source, and by comparing near and far neutrino energy spectra, one gains information about the oscillation probability.

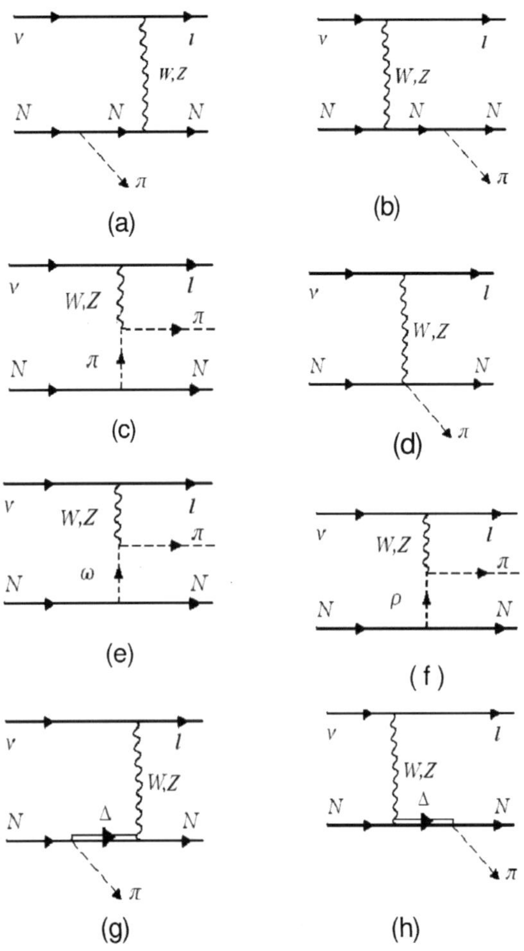

Figure 1. Different amplitude contributions to νN scattering

In this moment new high quality data are becoming available from MiniBoone [3] and SciBoone [4], experiments full dedicated to measure cross sections, which will sum to the already obtained in the K2K [5] experiment. A proper understanding of these processes is very important in the analysis of neutrino oscillation experiments. For instance, the π^0 production in NC is the most important background to experiments that measure $\nu_\mu \rightarrow \nu_e$ oscillation in the energy range around 1 GeV. This is because the emerging π^0 can mimic the ν_e signal events when, for example one of the two photons in the $\pi^0 \rightarrow \gamma\gamma$ decay (used to detect the pion) is not detected. This can happen when the photon exits the detector before showering or does not have enough energy to initiate the shower. Similarly π^+ production by CC is an important source of background in $\nu_\mu \rightarrow \nu_x$ disappearance searches.

In the energy region relevant for MiniBooNE and SciBoone experiments that is between 0.5 and 2 GEV, the dominant process in addition to the QE scattering is the excitation of the Δ (P_{33}(1232 MeV)) resonance (figure 1(g),(h)). We also have in addition many non resonant contributions (figure 1(a)-(f)) to the CC1π and NC1π cross sections. It is then important to develop a consistent treatment from the point of view of the effective Lagrangian models for describing the Δ (spin 3/2 particle). These very actual experiments, require to include medium effects on the free nucleon cross section to make a comparison between theoretical results and the data. Nevertheless and before to analyze these effects, it is important that the model used to describe the free nucleon cross section incorporates consistently the resonances and also non resonant terms, since the interference between them could be important. This model should be an extension of one used satisfactorily in describing other reactions to get a confiable theoretical determination of the 1π background cross section, already at the free nucleon level, to be useful at the moment of the calibration of the experiments. Here we use the axial form factors already obtained from the study of the CC1π production on free nucleons [6], within an isobar model we have developed to describe several other different processes. We will analyze its predictive power of the model on NC1π production of free nucleons.

2. $\Delta(1232)$ amplitude

Effective Lagrangians used to built non resonant Feynman graphs are rough well known and the corresponding amplitudes can be obtained straightforward [6], then we concentrate in those including the resonance terms

$$\hat{\mathcal{L}}_\Delta = \hat{\mathcal{L}}_{\Delta\text{free}} + \hat{\mathcal{L}}_{\Delta\pi N} + \hat{\mathcal{L}}_{\Delta WN},$$

where $\hat{\mathcal{L}}_{\Delta\text{free}}$ is the free Δ contribution, $\hat{\mathcal{L}}_{\Delta\pi N}$ accounts for the strong interaction and $\hat{\mathcal{L}}_{\Delta WN}$ the weak one. $\hat{\mathcal{L}}_\Delta$ depends in each of its pieces on an arbitrary parameter A through the matrix [7]

$$\Lambda_{\mu\nu}(A) = g_{\mu\nu} + \frac{1}{2}(1+3A)\gamma_\mu\gamma_\nu,$$

being invariant under the contact transformation

$$\psi^\mu_\Delta \to \psi^\mu_\Delta + a\gamma^\mu\gamma_\alpha\psi^\alpha_\Delta\,, \qquad A \to A' = \frac{A-2a}{1+4a},$$

where a ($a \neq -1/4$) is another arbitrary parameter. This field transformation assures that spurious spin-1/2 components ($\frac{1}{2}\bigotimes 1 = \frac{1}{2}\bigoplus\frac{3}{2}$) are removed from the field describing a free Δ particle. Physical amplitudes should be independent of this parameter and in what follows we give the set Feynman rules to built the contributions shown in figures 1(g) and 1(h). The effective weak Δ-excitation (V-A) vertex amplitude reads

$$\langle\Delta|V_\mu(0)|N\rangle = \sqrt{2}C^V_B\overline{u}_{\Delta\nu}\left[(G_M(Q^2) - G_E(Q^2))K^C_{\nu\mu}\right](\mathbf{T}^\dagger\cdot\mathbf{B}^*)u_N,$$

where $K^M_{\nu\mu}$, $K^E_{\nu\mu}$, and $K^C_{\nu\mu}$ are electromagnetic vertex functions [8], $\mathbf{B} \equiv \mathbf{W},\mathbf{Z}$ are isospin wavefunctions, $C^V_{W_\pm} = \sqrt{2}cos\theta$ and $C^V_Z = (1-2sin^2\theta_W)$, and

$$\begin{aligned}\langle\Delta|A_\mu(0)|N\rangle &= iC^A_B\overline{u}_{\Delta\nu}\Bigg[-D_1(Q^2)g_{\nu\mu} + \frac{D_2(Q^2)}{m_N^2}(\mathrm{p}+\mathrm{p}_\Delta)^\alpha \\ &\quad\times(g_{\nu\mu}\mathrm{q}_\alpha - \mathrm{q}_\nu g_{\alpha\mu}) - \frac{D_3(Q^2)}{m_N^2}\mathrm{p}_\nu\mathrm{q}_\mu\Bigg](\mathbf{T}^\dagger\cdot\mathbf{B}^*)u_N\end{aligned}$$

where $C^A_{W_\pm} = \sqrt{2}cos\theta$ and $C^A_Z = 1$, and where the Δ deformation has been not considered ($D_4 = 0$). The strong $\pi N\Delta$ formation vertex amplitude reads

$$\mathcal{V}(\pi N \to \Delta) = -\left(\frac{f_{\Delta\pi N}}{m_\pi}\right)q^\nu_\pi\overline{u}_{\Delta\nu}(\mathbf{T}^\dagger\cdot\pi^*)u_N,$$

Journal of Physics: Conference Series **239** (2010) 012010 doi:10.1088/1742-6596/239/1/012010

while the bare Δ propagator is

$$G^0_{\alpha\beta}(\mathrm{p}) = \frac{\not{\mathrm{p}}+m_\Delta}{\mathrm{p}^2-m_\Delta^2}\left\{-g_{\alpha\beta}+\frac{1}{3}\gamma_\alpha\gamma_\beta+\frac{1}{3m_\Delta}(\gamma_\alpha \mathrm{p}_\beta-\gamma_\beta \mathrm{p}_\alpha)+\frac{2}{3m_\Delta^2}\mathrm{p}_\alpha \mathrm{p}_\beta \right.$$
$$\left. -\frac{2(\mathrm{p}^2-m_\Delta^2)}{3m_\Delta^2}\left[\gamma_\alpha \mathrm{p}_\beta-\gamma_\beta \mathrm{p}_\alpha-(\not{\mathrm{p}}+m_\Delta)\gamma_\alpha\gamma_\beta\right]\right\}.$$

Note that if we drop the term in the second line of the previous equation we get the so called Rarita-Schwinger(RS) propagator. Several approaches use the vertexes shown above together the RS propagator, being this procedure inconsistent [9]. Until this moment we have introduced bare vertexes and propagator for the Δ, and they should be dressed by the πN interaction to all orders [10]. Numerical calculation of the dressed vertexes and propagator is a difficult task because as we are in presence of a non-renormalizable effective theory one must included cutting FF to calculate loop integrals, which introduce model dependencies. Nevertheless it has been shown that for $W \sim m_\Delta$, where W is the πN invariant mass, one can assume that a dressed vertex can be replaced for the bare one but with an *effective* (strong, electromagnetic or weak) coupling constant [10]. The unstable character of the resonance is included in the unperturbed propagator G^0_Δ by the replacement $G = G^0(m_\Delta \to m_\Delta - i\Gamma_\Delta/2)$,where m_Δ now is the dressed mass and Γ_Δ the resonance width, this independent-energy self-energy approximation is called the complex mass scheme [7].

In terms of the center of mass (CM) variables the total cross section for the $\nu N \to l\pi N'$ process reads ($E_\nu^{CM} = \frac{m_N E_\nu^{Lab}}{\sqrt{2E_\nu^{Lab} m_N + m_N^2}}$) reads [6]

$$\sigma(E_\nu^{CM}) = \frac{m_N^2 N_l}{(2\pi)^4 E_\nu^{CM}\sqrt{s}}\int_{E_l^-}^{E_l^+} dE_l^{CM}\int_{E_\pi^-}^{E_\pi^+} dE_\pi^{CM}\int_{-1}^{+1} dcos\theta$$
$$\int_0^{2\pi} d\eta \frac{1}{16}\sum_{spin}|\mathcal{M}|^2, \qquad (1)$$

where $\sqrt{s} = E_\nu^{CM} + E_N^{CM}$, and where $\mathcal{M} = \mathcal{M}_B + \mathcal{M}_R$ is separated in a background and resonant contribution. In the background amplitudes the vector FF are fixed through the CVC hypothesis, the axial ones using the PCAC hypotesis, and the strange FF obtained from parity violating electron scattering. Nevertheless the resonance parameters should be fixed by fitting cross sections calculated within our consistent isobar model (CIM), to different experimental data. For example the fitting to the elastic $\pi^+ p \to \pi^+ p$ cross section data, leads to $\mathrm{f}^2_{\Delta N\pi}/4\pi = 0.317 \pm 0.003$, $m_\Delta = 1211.7 \pm 0.4 MeV$ and $\Gamma_\Delta = 92.2 \pm 0.4 MeV$ [11]. We will assume the CVC hypothesis for the vector sector of the resonance FF. Then we can take them from a fit to the experimental data analysis for the $M_{1+}^{3/2}$, $E_{1+}^{3/2}$ amplitude multipoles, extracted from the $\gamma p \to \pi^0 p$ and $\gamma p \to \pi^+ n$ processes. We got the dressed values $G_M \equiv G_M(0) = 2.97 \pm 0.08$ and $G_E \equiv G_E(0) = 0.055 \pm 0.010$, and the bare values $G_M^0 = 1.69 \pm 0.02$ and $G_E^0 = 0.028 \pm 0.008$ through an evaluation of the loop integrals [10]. Finally the axial FF at $Q^2 = 0$, $D_i(0), i = 1, 3$, are obtained by comparing the non-relativistic limit of the previous matrix element in the Δ rest frame with the non-relativistic quark model, and all can be expressed in terms of $D_1(0)$ [12]. Making a fit to the flux averaged differential cross section $\frac{d<\sigma>}{dQ^2}$ in the CC1π $\nu p \to \mu^- p \pi^+$ reaction we get $D_1(0) = 2.35 \pm 0.05$, or as in the usual nomenclature $C_5^A(0) = 1.35 \pm 0.03$ [6].

3. Results for NC

We have four processes of interest in NC production of π

$$\nu_\mu p \rightarrow \nu_\mu p \pi^0,$$
$$\nu_\mu p \rightarrow \nu_\mu n \pi^+,$$
$$\nu_\mu n \rightarrow \nu_\mu n \pi^0,$$
$$\nu_\mu n \rightarrow \nu_\mu p \pi^-,$$

and data on single-pion production on nucleon targets via NC are scarce, old, and being the experiments usually done on complex nuclear targets. In spite of this, we intend a comparison with data coming from scattering on deuteron and with the results of a simulation devoted to extract neutrino-nucleon cross section from a neutrino-nucleus experiment. First we compare our results within the CIM model to the $\nu_\mu d \rightarrow \nu_\mu p \pi^- p_s$ process measured in the Argonne (ANL) 12 foot deuterim bubble chamber [13, 14] exposed to a broad neutrino beam from the Zero Gradient Synchroton (ZGS) with $0.3 \leq E_\nu \leq 1.5\ GeV$. The cut in the E_ν energy is to avoid double-pion production and p_s denotes the spectator proton in the reaction, and it is expected that due to the small binding energy of d ($\sim$ 2 MeV) one can compare the data with the theoretical calculations for $\nu_\mu n \rightarrow \nu_\mu p \pi^-$. This experiment only gives results for the flux averaged $\nu_\mu p \rightarrow \nu_\mu n \pi^+$ and $\nu_\mu p \rightarrow \nu_\mu p \pi^0$ NC cross sections.

In addition we have to our disposal data on NC reactions coming from simulations obtained with the NUANCE software package, which includes Fermi motion, Pauli blocking and final state interactions (FSI), to extract single-pion production on free nucleons from a nuclear target. A reanalysis was done in 2002 by E. Hawker [15] (NUANCE 2002) on the Gargamelle bubble chamber NC1π data. This detector is filled with propane-freon ($C_2H_8 + CF_3Br$) at CERN [16]. The NUANCE simulation code to get free cross section from the propane-freon target was used, since in this experiments only the integrated flux averaged cross section an the R-ratios between NC and CC cross sections were reported. No cuts in the neutrino energy was done and the flux runs on an interval $1GeV \leq E_\nu \leq 8GeV$.

In the figure 2 we compare theoretical results using eq. (1) within the CIM model for the $\nu_\mu n \rightarrow \nu_\mu p \pi^-$ reaction with the data of [13]. We make the cut $W \leq 1.45GeV$ in order to keep us within the first resonance region and to keep valid the complex mass scheme. Also we report the values for the R-ratios measured in the ANL experiment and our calculated values in the table 1. The $\sigma(\nu n \rightarrow \nu p \pi^-)$ prediction with the CIM is consistent with the experimental cross section in spite of the large errobars. The replacement of the full propagator by the RS one, leads to appreciable changes and the B-R interference is important. $R_\pm$ fall into the limits of the experiment and also the data are consistent with the fact that in our model both $\nu n \rightarrow \nu p \pi^-$ and $\nu p \rightarrow \nu n \pi^+$ amplitudes have the same isospin coefficients. Our calculated value for R_0 is out of the experiment limits, nevertheless it must be taken into account that being $\nu p \rightarrow \nu n \pi^+$ and $\nu p \rightarrow \nu p \pi^0$ measured as different channels of the same experiment, while $\nu n \pi^+$ events have a 30% of background events, $\nu p \pi^0$ have a 65% [14].

Now we show in the figure 3 results for the $\nu_\mu p \rightarrow \nu_\mu p \pi^0$ reaction not measured in ANL but reported in the CERN experiment, being its cross section extracted with the NUANCE code, and compare it with our prediction with the CIM model. CIM results are consistent with the

Table 1. Comparison of R-ratios for the reactions measured in [13, 14] with results obtained with the CIM model.

$R_- = \frac{<\sigma(\nu n \rightarrow \nu p \pi^-)>}{<\sigma(\nu p \rightarrow \mu^- p \pi^+)>}$	0.11± 0.022(Derrick80)	0.104(CIM)
$R_+ = \frac{<\sigma(\nu p \rightarrow \nu n \pi^+)>}{<\sigma(\nu p \rightarrow \mu^- p \pi^+)>}$	0.12± 0.04(Derrick81)	0.104(CIM)(isos. sym.)
$R_0 = \frac{<\sigma(\nu p \rightarrow \nu p \pi^0)>}{<\sigma(\nu p \rightarrow \mu^- p \pi^+)>}$	0.09± 0.05(Derrick81)	0.207(CIM)

Journal of Physics: Conference Series **239** (2010) 012010 doi:10.1088/1742-6596/239/1/012010

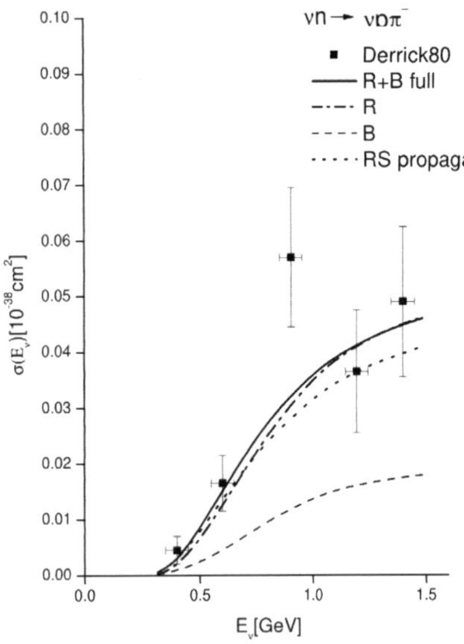

Figure 2. Theoretical calculations for the $\nu_\mu n \to \nu_\mu p\pi^-$ reaction are compared with the data of [13]. With B we indicate non resonant contributions plus the 1(g) one of the figure 1, and while with R only the pole contribution of the graph 1(h). We also show the full amplitude obtained with the RS propagator.

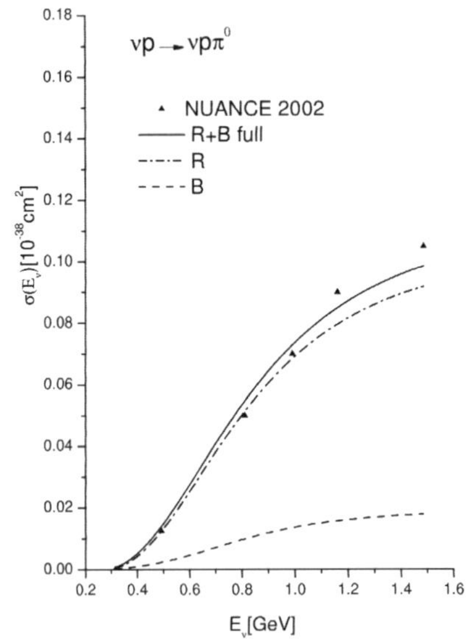

Figure 3. Theoretical calculations for the $\nu_\mu p \to \nu_\mu p\pi^0$ reaction are compared with the simulation of [15]. Lines convention is the same as in figure 2.

NUANCE simulation for $\sigma(\nu_\mu p \to \nu_\mu p\pi^0)$ cross section in spite of the W cut in our calculations, which could be understood examining the $d\sigma/dW$ distribution for the CERN experiment, which shows a not appreciable number of events in the $\nu_\mu p\pi^0$ channel for $W > 1.45GeV$. Finally we mention that in CERN experience they report

$$R_0 = \frac{<\sigma(\nu p \to \nu p\pi^0)> + <\sigma(\nu n \to \nu n\pi^0)>}{2 < \sigma(\nu n \to \mu^- p\pi^0)>} = 0.45 \pm 0.08,$$

in comparison with out theoretical value $R_0 = 0.766$. The lower value of R_0 regards our calculation could be due to our W cut, but as is shown in the NUANCE simulation the measured value for $\sigma(\nu n \to \nu n\pi^0)$ (equal to $\sigma(\nu p \to \nu p\pi^0)$ within the CIM) is critically low, which could be due to the challenging final state to detect.

4. Conclusions

In summary, we have proved the predictive power of our consistent dynamical model also for weak NC1π pion production. The model was previously used to describe pion-nucleon elastic and radiative scattering, pion photoproduction and CC1π in a satisfactory way. We show, as in our previous calculations, the importance of using the full Δ propagator, which is consistent with the $N \to \Delta$ weak vertex. Also we again mention the importance of the interference between the resonant and background contributions in the weak amplitude.

Actual Miniboone results [3] correspond to an energy region $0.4 < E_\nu < 2.4GeV$ and no invariant mass cut is used, events occurring in the region $W < 1.6GeV$, which is above the Δ one. Then, it is imperative to add consistently to the resonant amplitude the second resonance region ($N(1440)P_{11}$, $N(1520)D_{13}$, $N(1535)S_{11}$,etc.) contributions. Also we need to incorporate energy dependent contributions to the Δ self-energy but keeping invariance on contact transformations of the Lagrangian. Finally medium effects should be incorporated in a way this invariance and also electromagnetic gauge invariance are not violated.

Acknowledgments

A. M. and C. B. fellow to CONICET (Argentina), and this work was supported by the UNLP under the grant for visits 2009. G.L.C. acknowledges partial support from UNLP, under grant for visits 2008.

References

[1] Gran R 2007 *Minerva Collaboration Preprint* hep-ex/07113029
[2] Raaf J L 2005 *Nuc. Phys. Proc. Suppl.* B **139** 47
[3] Aguilar-Arevalo A A *et al* 2009 *Miniboone Collaboration Preprint* hep-ex/0904.3159
[4] Hiraide K 2009 *Sciboone Collaboration Preprint* hep-ex/0909.5127
[5] Hasegawa M *et al* 2008 *K2K Collaboration Phys. Rev. Lett.* **95** 252301
[6] Barbero C, López Castro G and Mariano A 2008 *Phys. Lett.* B **664** 70
[7] El-Amiri M, López Castro G and Pestieau J 1992 *Nucl. Phys.* A **543** 673
[8] Jones H F and Scadron M D 1973 *Ann. Phys.* **81** 1
[9] Hernandez E, Nieves J and Valverde M 2007 *Phys. Rev.* D **76** 033005
[10] Mariano A 2007 *J. Phys. G: Nucl. Part. Phys.* **34** 1627
[11] López Castro G and Mariano A 2001 *Phys. Lett.* B **517** 339
[12] Sato T, Uno D and Lee T S H 2002 *Phys. Rev.* C **67** 065201
[13] Derrick M *et al* 1980 *Phys. Lett.* B **92** 363
[14] Derrick M *et al* 1981 *Phys. Rev.* D **23** 569
[15] Hawker E A 2002 *Proceedings of the Second International Workshop on Neutrino-Nucleus Interactions in the Few GeV Region* http://www.ps.uci.edu/ñuint/proceedings/hawker.pdf
[16] Krenz W *et al* 1978 *Nuc. Phys.* B **135** 45

Exploring the Nucleon Structure from First Principles of QCD

W Bietenholz[1], N Cundy[2], M Göckeler[2], R Horsley[3], H Perlt[4], D. Pleiter[5], P E L Rakow[6], G Schierholz[2,7], A Schiller[4], T Streuer[2] and J M Zanotti[3] (QCDSF Collaboration)

[1] Instituto de Ciencias Nucleares, Universidad Nacional Autónoma de México, A.P. 70-543, C.P. 04510 México, Distrito Federal, Mexico
[2] Institut für Theoretische Physik, Universität Regensburg, 93040 Regensburg, Germany
[3] School of Physics, University of Edinburgh, Edinburgh EH9 3JZ, United Kingdom
[4] Institut für Theoretische Physik, Universität Leipzig, 04109 Leipzig, Germany
[5] John von Neumann Institut für Computing, Deutsches Elektronen-Synchrotron DESY, 15738 Zeuthen, Germany
[6] Theoretical Physics Division, Dept. of Mathematical Sciences, University of Liverpool, Liverpool, L69 3BX, United Kingdom
[7] Deutsches Elektronen-Synchrotron DESY, 22603 Hamburg, Germany

E-mail: wolbi@nucleares.unam.mx

Abstract. Quantum Chromodynamics (QCD) is generally assumed to be the fundamental theory underlying nuclear physics. In recent years there is progress towards investigating the nucleon structure from first principles of QCD. Although this structure is best revealed in Deep Inelastic Scattering, a consistent analysis has to be performed in a fully non-perturbative scheme. The only known method for this purpose are lattice simulations. We first sketch the ideas of Monte Carlo simulations in lattice gauge theory. Then we comment in particular on the issues of chiral symmetry and operator mixing. Finally we present our results for the Bjorken variable of a single quark, and for the second Nachtmann moment of the nucleon structure functions.

1. Introduction

For 37 years we have a strong candidate for a fundamental theory underlying nuclear physics. There is a well-founded paradigm that the whole of nuclear physics can in principle be explained by QCD. Nevertheless this theory has hardly been represented at this workshop, because most talks referred to complex phenomena, which can in practice still not be computed based on QCD. Effective approaches — like the shell-model — are popular for this purpose. They are often successful for specific questions, but they all have their limitations.

In this contribution we do address QCD, and we report in particular on progress in the understanding of the nucleon structure function from first principles. This is obviously a non-perturbative task, and the only method to tackle it systematically are lattice simulations.

The formulation of QCD looks amazingly simple: it deals with quarks with a local $SU(3)$ symmetry ("colour symmetry"), which gives rise to strong interaction mediated by 8 gauge bosons ("gluons"); one for each generator of the gauge group $SU(3)$. Since the latter is non-Abelian, the gluons also interact among each other, which makes the dynamics of the system complicated. In the low energy regime, which dominates our daily experience, the gauge coupling is strong, and the interactions are extremely involved. This can be seen for instance from the fact that the fundamental masses of the lightest (valence) quark flavours — which we assume to be generated by the Higgs mechanism — only contribute about 2 % to the nucleon mass; all the rest is contained in some dense "foam" of gluons and sea-quarks.

So far perturbation theory could be carried out for some quantities up to α_s^4, with $\alpha_s = g_s^2/4\pi$, g_s being to strong gauge coupling. This is applicable to a number of high energy processes, dominated by the exchange of few gluons. A suitable method for that purpose is dimensional regularisation, which is, however, strictly limited to perturbation theory.

On the other hand, the lattice regularisation enables a fully non-perturbative treatment, *i.e.* we can really address finite values of α_s. Here we refer to the functional integral formulation of quantum field theory, which is (in principle) based on an integration over all possible configurations of the fields involved. The partition function, and all n-point functions that can be derived from it, are then given by an integration over all these configurations with a phase factor $\exp(\mathrm{i}S)$, where S is the action of the configuration. The convergence of such integrals can be improved drastically by moving to Euclidean space-time, *i.e.* by invoking an "imaginary time" (Wick rotation). This changes the factor mentioned above to $\exp(-S_{\rm E})$, where $S_{\rm E}$ is the Euclidean action. This transition is justified in view of the expectation values of n-point functions if four conditions — known as the "Osterwalder-Schrader axioms" — hold. For many theories, including QCD at zero vacuum angle θ and zero baryon density, $S_{\rm E}$ is real positive for all configurations, hence the factor $\exp(-S_{\rm E})$ can be interpreted as a probability distribution for the possible field configurations.

Lattice studies consider discrete points in a finite volume of Euclidean space-time. The matter fields — in this case the quark fields $\bar{\Psi}$, Ψ — are located on these lattice sites, while the gauge field variables $U_\mu \in SU(3)$ live on the links connecting nearest neighbour sites. Thus the functional integral is completely well-defined. Most theories of interest are bilinear in the fermion fields. Then the action takes the structure $S_{\rm E}[\bar{\Psi}, \Psi, U] = \bar{\Psi} M[U] \Psi + S_{\rm gauge}[U]$, where the indices of $\bar{\Psi}$, Ψ run over all lattice sites, and on each site over all internal degrees of freedom. The fermion matrix M contains the gauge couplings between neighbouring sites (due to the discrete covariant derivative in the Dirac operator). $S_{\rm gauge}$ is the pure gauge part of the action. In accordance with the Spin-Statistics Theorem, the components of the spinor fields $\bar{\Psi}$, Ψ are given by Grassmann variables, *i.e.* they anti-commute, and the corresponding integration yields

$$Z = \int \mathcal{D}\bar{\Psi}\mathcal{D}\Psi\mathcal{D}U \, \exp(-S_{\rm E}[\bar{\Psi}, \Psi, U]) = \int \mathcal{D}U \, \det M[U] \, e^{-S_{\rm gauge}[U]} \, . \tag{1}$$

The functional measure $\mathcal{D}U$ now represents the integral over the (compact, gauge invariant) Haar measure for each link variable. Thus we preserve exact gauge invariance on the regularised level. For system sizes of interest, it is still far beyond our ability to compute this integral explicitly. Hence the method of choice is to generate a large set of gauge configurations with the given probability distribution; moreover the configurations should be well de-correlated, *i.e.* statistically independent from each other. There are powerful algorithms for this purpose, see Refs. [1] for reviews. Once such a set of configurations is generated, it can be used to measure observables of interest. This approximate functional integral yields results for expectation values with statistical errors (due to the finite sample of configurations) and systematic errors (finite lattice spacing, finite volume etc.). These errors can be estimated and reduced with additional computational effort. The important point, however, is that the result is fully non-perturbative. This allows us nowadays for instance to calculate the light hadron spectrum from first principles of QCD up to a precision of a few percent. The results agree with phenomenology within these errors.

Our daily experience is dominated by nuclear physics at low energy. Regarding QCD this implies a dominance by the light quark flavours, in particular the flavours u and d. Their masses are very small compared to $\Lambda_{\rm QCD}$ — the intrinsic scale of of QCD, which provides a scale for the hadron masses. Hence these quarks have an approximate chiral symmetry, which is of prominent importance in low energy QCD. To briefly review this property, we first note that fermion fields can be decomposed into "left- and right-handed" components by means of the chiral projectors,

$$\Psi_{\rm L,R} = \frac{1}{2}(1 \pm \gamma_5)\,\Psi\ , \quad \bar\Psi_{\rm L,R} = \bar\Psi\,\frac{1}{2}(1 \mp \gamma_5)\ . \tag{2}$$

In the chiral limit (zero quark masses) these two components decouple. It is allowed to add quark masses to the Lagrangian, since QCD is a vector theory — the left- and right-handed components couple in the same way to the gauge field — (unlike the electroweak sector of the Standard Model). From a perturbative perspective, the approximate chiral symmetry still protects the masses of light quarks from large renormalisation. Based on this property, it is occasionally claimed that there is no hierarchy problem for fermion masses. However, it is the non-perturbative level that ultimately matters for physics, and there it is highly non-trivial to implement (approximate) chiral symmetry in a regularised system. A quite obvious ansatz is the Wilson lattice fermion, which starts from the naïve discretisation of the derivative in the Dirac operator, and subtracts a discrete Laplacian to send the fermion doublers to the cutoff scale. However, this extra term breaks chiral symmetry explicitly. As a consequence, under gauge interaction the fermion masses are renormalised such that they naturally end up at the cutoff scale as well. The (observed) property $m_{u,d} \ll \Lambda_{\rm QCD}$ can then only be attained by a tedious fine-tuning of negative bare quark masses, *i.e.* the hierarchy problem is back.

Only at the very end of the last century was there a breakthrough in the formulation of chiral lattice fermions, which overcomes this problem, at least in vector theories. A key observation was that a lattice modified version of chiral symmetry of the form [2]

$$\bar\Psi \to \bar\Psi\Big(1+\varepsilon\Big[1-\frac{a}{2}D\Big]\gamma_5\Big)\ , \quad \Psi \to \Big(1+\varepsilon\gamma_5\Big[1-\frac{a}{2}D\Big]\Big)\Psi \qquad (\text{to } O(\varepsilon))\ , \tag{3}$$

is sufficient to rule out additive mass renormalisation. Here a is the lattice spacing, and in the continuum limit this turns into the standard chiral transformation (infinitesimal form). Even on the lattice the fermionic Lagrangian $\bar\Psi D\Psi$ is invariant, *if* the lattice Dirac operator D fulfils the Ginsparg-Wilson relation (GWR)

$$\{D,\gamma_5\} \equiv D\gamma_5 + \gamma_5 D = aD\gamma_5 D\ . \tag{4}$$

On the other hand, the functional measure $\mathcal{D}\bar\Psi\mathcal{D}\Psi$ is not invariant, and this is just what it takes to reproduce the axial anomaly correctly [2].

Solutions to the GWR can be constructed based on block spinor renormalisation group transformations ("perfect lattice fermion") [3, 4], or by pulling apart the chiral modes in an extra "dimension" ("Domain Wall Fermion") [5]. By integrating out that "dimension", one obtains — for γ_5-Hermitian operators, $D^\dagger = \gamma_5 D\gamma_5$ — the "overlap Dirac operator" $D_{\rm ov}$ [6],

$$aD_{\rm ov} = 1 + A_{\rm ov}\ , \quad A_{\rm ov} = A/\sqrt{A^\dagger A}\ , \quad A = aD - 1\ . \tag{5}$$

It is easy to verify that it satisfies the GWR, which can now be written as $D_{\rm ov} + D_{\rm ov}^\dagger = aD_{\rm ov}^\dagger D_{\rm ov}$. For the kernel D we can insert for instance the Wilson operator [6], but properties like scaling, locality and approximate rotation symmetry can be improved by using a better kernel [7].

In addition to the exact, lattice modified chiral symmetry, the GWR also prevents uncontrolled operator mixing on the regularised level [4]. This powerful property is very helpful, for instance for the numerical study of a fully non-perturbative Operator Product Expansion (OPE).

We now summarise our lattice OPE project [8, 9, 10] as an example for the application of chiral quarks in lattice QCD. This takes us beyond the hadron spectrum; we investigate the internal structure of the nucleon (see Ref. [11] for a recent review). Phenomenologically this structure is revealed in Deep Inelastic Scattering of a lepton on a nucleon target. This scattering is dominated by a one-photon exchange involving a single quark, as sketched below.

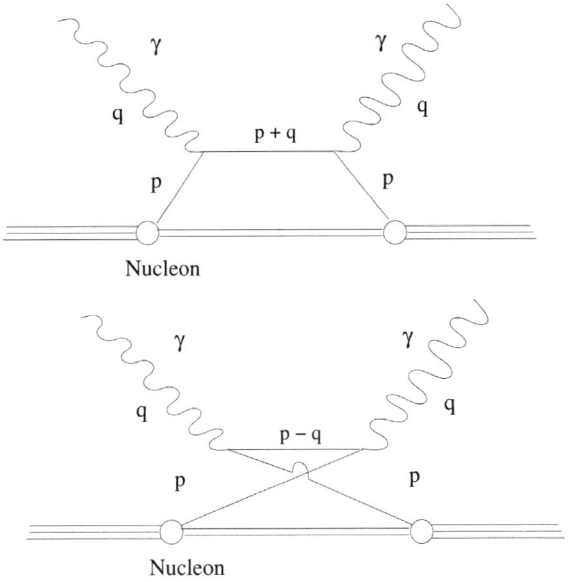

Although this is a high energy process, there are problems with a perturbative treatment due to ambiguous UV and IR divergences. A fully satisfactory study based on QCD has to be consistently non-perturbative, *i.e.* it is a challenge for the lattice. The involvement of a single quark implies that chirality is conserved, hence the use of Ginsparg-Wilson quarks is appropriate. In our simulations we applied 2 flavours of degenerate overlap valence quarks with masses of 29 MeV and 73 MeV, which correspond to pion masses of 280 MeV and 440 MeV, respectively. The goal is the evaluation of the product of electromagnetic currents J_μ — of the photon exchange, with large transfer momentum q — between quark states of low momentum p. This product is decomposed by the OPE as follows,

$$W_{\mu\nu} \simeq \langle \Psi(p)|J_\mu(q)J_\nu(-q)|\Psi(p)\rangle = \sum_{m,i,n} C^{(m)}_{\mu\nu,i,\mu_1\dots\mu_n}(q)\langle\Psi(p)|\mathcal{O}^{(m)}_{i,\mu_1\dots\mu_n}|\Psi(p)\rangle \ . \tag{6}$$

$C^{(m)}$ are Wilson coefficients, which are independent of the target and therefore of p, and $\mathcal{O}^{(m)}$ are *local* operators, which are relevant to describe the nucleon structure. They involve the quark momentum components p_{μ_j}. We further use the Clifford index $i = 1\dots 16$, and the index m to distinguish operators with the same symmetries.

A truncation of the OPE on the right-hand-side of eq. (6) requires the scale separation

$$p^2 \ll q^2 \ . \tag{7}$$

We probed three photon momenta, $|q^{(1)}| = 2.2$ GeV, $|q^{(2)}| = 3.3$ GeV, $|q^{(3)}| = 4.4$ GeV. Based on the above scale separation, we truncated the quark bilinears at $O(|p|^3)$. Thus we still include 1360 operators, but for photon momenta $q \propto (1,1,1,1)$ they occur in only 67 equivalence classes. The strategy is now to measure the elements of $W_{\mu\nu}$ and $\langle\mathcal{O}^{(m)}\rangle$ for a multitude of low quark

Journal of Physics: Conference Series **239** (2010) 012011 doi:10.1088/1742-6596/239/1/012011

momenta p, and solve eq. (6) for the Wilson coefficients. They are over-determined in this way, but by means of a Singular Value Decomposition we arrived at reliable values [8]. In Fig. 1 we compare the results obtained numerically at photon momentum $q^{(2)}$ to the Wilson coefficients at tree level in the chiral limit. We see that they follow the same pattern, though gauge interaction reduces the absolute values significantly. In particular the coefficients C_1 and $C_7 \dots C_{16}$ vanish in the chiral limit, because they are attached to operators with an even power of γ-matrices. We obtain small absolute values for them also in the interacting case at finite quark mass, which confirms that chiral symmetry on the lattice works, and it suppresses operator mixing.

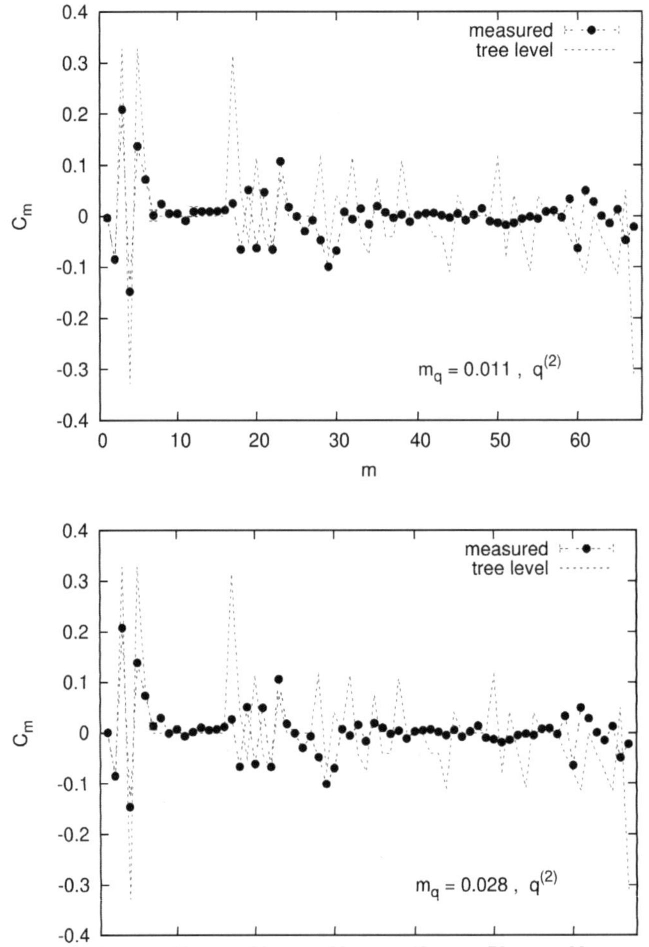

Figure 1. The Wilson coefficients for quark masses $m_q = 0.011$ and $m_q = 0.028$ in lattice units, corresponding to 29 MeV and 73 MeV. In both cases the photon momentum amounts to 3.3 GeV. The results for the 67 independent coefficients are compared to their values on tree level. We see a similar pattern, with reduced absolute values under interaction. The coefficients which vanish in the chiral limit, C_1 and $C_7 \dots C_{16}$, are still small at finite quark mass, which confirms the validity of our method.

As a first application, we evaluate the Bjorken variable x for one unpolarised quark. The expectation value $\langle x \rangle$ measures the nucleon momentum fraction carried by the quark involved in the Deep Inelastic Scattering. To be explicit, we consider the tensor [9]

$$T_{\mu\nu} = \frac{1}{4}\mathrm{Tr}[S^{-1}(p)W_{\mu\nu}(p,q)] = \delta_{\mu\nu}W_1 + p_\mu p_\nu W_2 + (p_\mu q_\nu + p_\nu q_\mu)W_4 + q_\mu q_\nu W_5 \ , \tag{8}$$

where S is the quark propagator. The Dirac-index structure is not visible here due to the trace (terms with W_3 and W_6 would occur for polarised targets).

In particular $T_{\mu\mu}$ is directly related to the Bjorken variable in the high energy Bjorken limit. In this limit, the Minkowski space momenta obey $(p \cdot q)^2 \gg p^2 q^2$. Among the 22 operators in the expansion of $T_{\mu\mu}$ we focus on those with spin 2, where the Bjorken limit leads to [9]

$$\langle x \rangle = \frac{q^2 \, [T_{\mu\mu}]_{\mathrm{spin}\ 2}}{2\sum_{\mu>\nu} p_\mu p_\nu} \ . \tag{9}$$

Fig. 2 (upper panel) shows the numerator against the denominator of the right-hand-side in eq. (9). A broad range of points measured at $m_q = 73$ MeV is consistent with a linear behaviour, which allows us to read off $\langle x \rangle$ from the slope (the errors are statistical only),

$$\langle x \rangle|_{q^{(1)}} = 0.44(9) \ , \quad \langle x \rangle|_{q^{(2)}} = 0.73(5) \ , \quad \langle x \rangle|_{q^{(3)}} = 0.60(5) \ . \tag{10}$$

where the errors are statistical only.

Finally we proceed to the nucleon structure function [10]. On a lattice of spacing a, a generic moment can be expanded as

$$\mathcal{M}(q^2) = c_2(q^2a^2)A_2(a^2) + \frac{c_4(q^2a^2)}{q^2}A_4(a^2) + \dots \tag{11}$$

where A_n are reduced nucleon matrix elements (the Lorentz structure is factored out), c_n are the corresponding Wilson coefficients, and n is the twist. Both factors are bare quantities. If one tries to replace a by some scale μ of a soft renormalisation scheme, one runs into ambiguities, which do not occur on the lattice with its hard momentum cutoff π/a.

In our study, we evaluated the (spin averaged) nucleon matrix elements for all 67 types of operators, and extracted moments of the structure function by Nachtmann integration of the hadronic tensor [12]. This allows us in particular to project out operators of spin 2. For instance, the traceless form of $W_{\mu\nu}$ leads to

$$\mathcal{M}(q^2) = \frac{3q^2}{(4\pi)^2}\int d\Omega_q \, n_\mu \Big(W_{\mu\nu} - \frac{1}{4}\delta_{\mu\nu}W_{\lambda\lambda}\Big) n_\nu \ \to \ \int_0^1 dx \, \Big(F_2(x,q^2) + \frac{1}{6}F_L(x,q^2)\Big) \ . \tag{12}$$

Here n is a constant 4-vector with $n^2 = 1$, and the last expression in eq. (12) refers again to the Bjorken limit. By means of a further projection [10] we can separate the structure functions F_2 and F_L. Based on the numerical measurement of the matrix elements, we can carry out the first integral in eq. (12). Thus we arrive at explicit and non-perturbative results for $\mathcal{M}$.

Our results for the second moment $\mathcal{M}_2$ are shown in Fig. 2 (lower panel). They agree within the errors for both quark masses, at all three photon momenta, which suggests that they may be valid up to the physical pion mass. As we increase the photon momentum (*i.e.* the transfer momentum in the hard scattering), the twist 4 contribution tends to vanish, while the twist 2 term stabilises at a moderate value. As usual the numerical errors increase for lighter quarks.

These results demonstrate that the explicit determination of the nucleon structure functions — as observed in Deep Inelastic Scattering — directly from QCD *is* feasible, in the framework of a non-perturbative regularisation free of ambiguities. Moreover we obtained explicit results for the Bjorken variable of the quark involved in this scattering. Our results provide insight into the nucleon structure. For a detailed discussion we refer to Ref. [10].

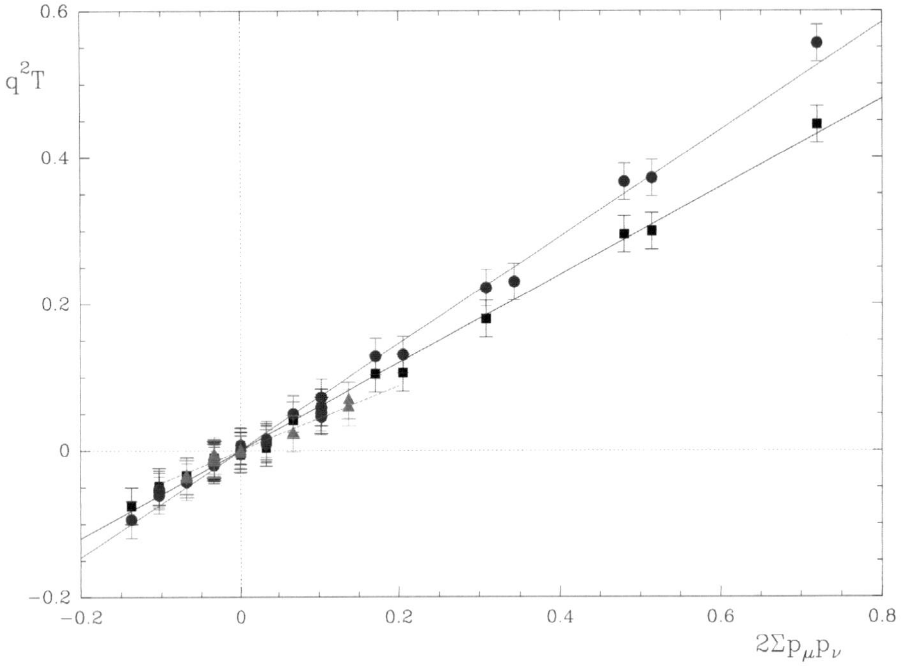

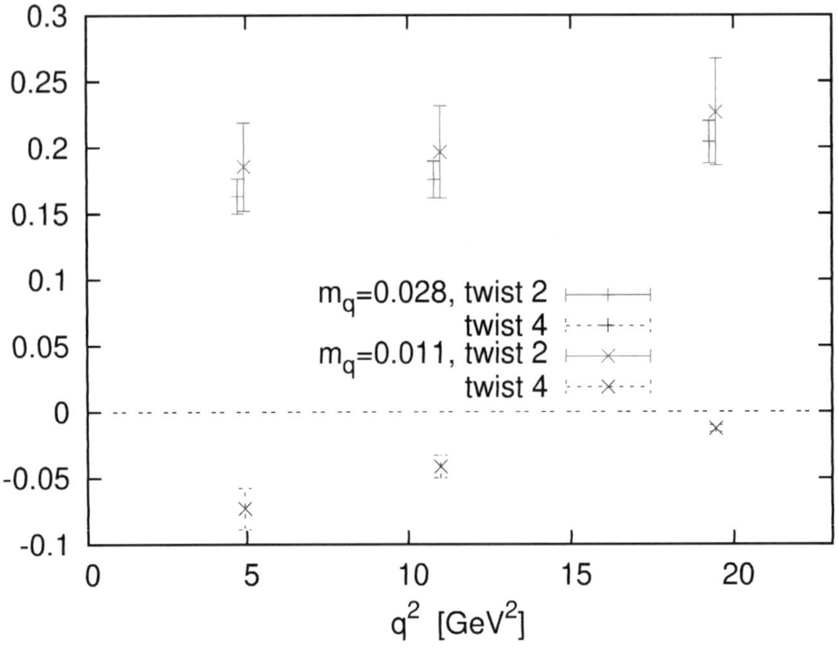

Figure 2. (Upper panel) The term $q^2 T_{\mu\mu}$ for spin 2, versus $2 \sum_{\mu>\nu} p_\mu p_\nu$ (numerator and denominator in eq. (9)). The squares, circles and triangles refer to photon momentum $q^{(1)}$, $q^{(2)}$, $q^{(3)}$, respectively. In all three cases the data points are compatible with a linear behaviour. Thus the slope allows for the determination of the Bjorken variables $\langle x \rangle$ given eq. (10). (Lower panel) The nucleon moment $\mathcal{M}_2$ against the photon momentum squared. We distinguish the contributions of twist 2 and twist 4. In all cases the result for our two quark masses agree within statistical errors. For large q^2 the twist 4 contribution seems to vanish, whereas the twist 2 contribution stabilises at a finite value in the range $\approx 0.5 \ldots 0.8$.

Acknowledgments
W.B. thanks the organisers of the interesting and pleasant workshop in Cocoyoc. This work was supported by the *Deutsche Forschungsgemeinschaft* (DFG) through project FOR 465 and *Sonderforschungsbereich* SFB/TR55. The computations were performed on clusters of the *Norddeutscher Verbund für Hoch- und Höchstleistungsrechnen* (HLRN).

References
[1] Kennedy AD 2006 (*preprint* hep-lat/0607038)
Lüscher M 2010 (*preprint* hep-lat/1002.4232)
[2] Lüscher M 1998 *Phys. Lett.* B **428** 342
[3] Ginsparg PH and Wilson KG 1982 *Phys. Rev.* D **25** 2649
Bietenholz W and Wiese U-J 1996 *Nucl. Phys.* B **464** 319
[4] Hasenfratz P 1998 *Nucl. Phys.* B **525** 401
[5] Kaplan D 1992 *Phys. Lett.* B **288** 342
[6] Neuberger H 1998 *Phys. Lett.* B **417** 141
1998 *Phys. Lett.* B **427** 353
[7] Bietenholz W 1999 *Eur. Phys. J.* C **6** 537
[8] Bietenholz W, Cundy N, Göckeler M, Horsley R, Perlt H, Pleiter D, Rakow PEL, Schierholz G, Schiller A, Streuer T and Zanotti JM 2009 *PoS(LAT2009)138* (*preprint* hep-lat/0910.2437)
[9] Bietenholz W, Cundy N, Göckeler M, Horsley R, Perlt H, Pleiter D, Rakow PEL, Schierholz G, Schiller A, Streuer T and Zanotti JM 2009 *PoS(LAT2009)139* (*preprint* hep-lat/0911.4892)
[10] QCDSF Collaboration, in preparation
[11] Hägler Ph 2009 (*preprint* hep-lat/0912.5483)
[12] Nachtmann O 1973 *Nucl. Phys.* B **63** 237

Journal of Physics: Conference Series **239** (2010) 012012 doi:10.1088/1742-6596/239/1/012012

NPDGamma: a measurement of the parity-violating gamma asymmetry A_γ in n-p capture

R C Gillis[1,†], R Alarcon[2], S Balascuta[2], L Barrón-Palos[3], J D Bowman[4,⋆], R D Carlini[5], W Chen[1], T E Chupp[6], C Crawford[7], S Covrig[5], M Dabaghyan[8], N Fomin[9], S J Freedman[10], T R Gentile[11], M T Gericke[12], G L Greene[9,4], F W Hersman[8], T Ino[14], G L Jones[15], B Lauss[16], M Leuschner[1], W R Lozowski[1], R Mahurin[5,12], Y Masuda[14], J Mei[1], G S Mitchell[17], S Muto[14], H Nann[1], S A Page[12], S I Penttilä[4], W D Ramsay[18], A Salas-Bacci[19], S Santra[1,a], M Sharma[6], P-N Seo[20], E I Sharapov[21], T B Smith[22], W M Snow[1], W S Wilburn[19], V Yuan[19]

[1] Indiana University, Bloomington, IN 47405, USA
[2] Arizona State University, Tempe, AZ 85287, USA
[3] Universidad Nacional Autónoma de México, C.P. 04510, D.F., Mexico
[4] Oak Ridge National Laboratory, Oak Ridge, TN 37831, USA
[5] Thomas Jefferson National Accelerator Facility, Newport News, VA 23606, USA
[6] University of Michigan, Ann Arbor, MI 48104, USA
[7] University of Kentucky, Lexington, Kentucky 40506, USA
[8] University of New Hampshire, Durham, NH 03824, USA
[9] University of Tennessee, Knoxville, TN 37996, USA
[10] University of California, Berkeley, CA 94720-7300, USA
[11] National Institute of Standards and Technology, Gaithersburg, MD 20899-0001,USA
[12] University of Manitoba, Winnipeg, MB, R3T 2N2, Canada
[14] High Energy Accelerator Research Organization, Oho 1-1, Tsukuba-shi, 305-0801, Japan
[15] Hamilton College, Clinton, NY 13323, USA
[16] Paul Scherrer Institut, CH-5232 Villigen, Switzerland
[17] Department of Biomedical Engineering, University of California, Davis, CA 95616, USA
[18] TRIUMF, 4004 Wesbrook Mall, Vancouver, BC, V6T 2A3, Canada
[19] Los Alamos National Laboratory, Los Alamos, NM 87545, USA
[20] Department of Physics, North Carolina State University, Raleigh, NC 27695, USA
[21] Joint Institute for Nuclear Research, Dubna, Russia
[22] University of Dayton, Dayton, OH 45469, USA
[a] Current address: Bhabha Atomic Research Center, Trombay, Mumbai 400085, India
⋆ Principal investigator

E-mail: † rcgillis@indiana.edu

Abstract. The NPDGamma Experiment measures the parity-violating correlation A_γ between neutron spin and photon momentum in the reaction $\vec{n} + p \to d + \gamma$. Knowledge of A_γ and other parity-violating observables in few-body nuclear systems will provide constraints for a parameterized description of $\Delta S = 0$ parity-violating phenomena free from complications of nuclear structure. The NPDGamma experiment uses a polarized cold pulsed neutron beam, a liquid parahydrogen target, and a cylindrical array of 48 CsI(Tl) scintillation detectors operated in current mode to search for the asymmetry. NPDGamma recently completed the first phase of the program to measure A_γ at the Los Alamos Neutron Science Center with the preliminary result $A_\gamma = (-1.2 \pm 2.1(\text{stat.}) \pm 0.1(\text{sys.})) \times 10^{-7}$, reproducing the previous upper limit from a measurement at a reactor facility. We discuss the theoretical background and experimental method and report on preliminary analysis of the LANSCE data. The second phase of the

Journal of Physics: Conference Series **239** (2010) 012012 doi:10.1088/1742-6596/239/1/012012

program to measure A_γ is in progress at the Spallation Neutron Source at Oak Ridge National Laboratory.

1. An introduction to the theory

Measurements of parity violation (PV) in the nucleon-nucleon (NN) system at low energies can provide unique insight into the physics of QCD in the strongly interacting limit. Although PV comes from the weak interaction, its manifestation in hadronic systems at low energies necessarily involves the strong interaction, with nonperturbative QCD complicating the calculation of the observables. The weak interaction is in general hard to isolate experimentally at low energies in the NN system due to the presence of the dominant strong and electromagnetic interactions. While large nuclei provide mechanisms that enhance PV observables — enough that many experimental observations have resolved PV effects — the interpretation of such processes often calls for an *ab initio* understanding of many-body systems that does not yet exist. Experiments now emphasize few-nucleon systems which can possess a theoretically-clean interpretation. Recent reviews on this topic are found in [1, 2, 3].

PV in the NN system is traditionally described in terms of the DDH model, a quark model calculation that uses symmetry arguments and information from nonleptonic hyperon decays to deduce estimates of PV couplings in a single-meson-exchange theory [4, 5]. This approach is similar to successful meson-exchange descriptions of the parity-conserving (PC) NN interaction. In the PV theory the three lightest mesons are used, with neutral pseudoscalar mesons excluded since in their case P violation also entails CP violation and such amplitudes are at best highly suppressed. One vertex of the meson exchange violates parity and corresponds to the coupling that is unique to the DDH theory. The other coupling is determined from measurements of PC observables. The PV couplings are labeled by $h_\mu^{\Delta I}$ where $\Delta I = 0, 1$ or 2 is the isospin change, and μ is the meson type. There are a total of six couplings: h_π^1, h_ρ^0, h_ρ^1, h_ρ^2, h_ω^0 and h_ω^1 (a seventh coupling $h_\rho^{1\prime}$ is considered negligible [6]). To reflect the considerable strong interaction uncertainities involved, DDH presented their work in the form of a set of best values with corresponding reasonable ranges.

More recently an effective field theory (EFT) approach is being developed for the PV NN interaction for the low energies and momenta relevant for most experiments [3, 7, 8, 9, 10]. The EFT approach is constructed to take into account all of the relevant symmetries and degrees of freedom in low energy QCD, in particular chiral symmetry. Since the perturbative expansion is in terms of the ratio of momentum to the QCD scale Λ, the unknown physics is hidden in the vertices. The theory then depends on parameters that can be determined from experiment. It is in principle possible to calculate the EFT parameters, however the physical insight associated with a model-based explanation is not as immediately apparent with the EFT approach.

The EFT approach has been expressed in different forms: one with pions fully integrated out, relevant for lower energies, and others with dynamical pions, including a possibility for two-pion exchange. The parameters of the pionless theory reduce at sufficiently low energies to a total of five independent S $\leftrightarrow$ P transition amplitudes. In the pionful theories, the same five parameters remain but incorporate different dynamics, and extra parameters that explicitly incorporate the pion dynamics are introduced.

The observable of interest in this experiment is the coefficient $A_{\gamma,PV}^{\vec{n}p}$ (A_γ when there is no ambiguity) of the PV correlation between neutron spin $\mathbf{s_n}$ and gamma momentum $\mathbf{k}_\gamma$ in the distribution $d\omega/d\Omega$ of gamma radiation from the capture reaction $\vec{n} + p \to d + \gamma$ (2.2 MeV):

$$\frac{d\omega}{d\Omega} \propto 1 + A_\gamma \frac{\mathbf{s_n} \cdot \mathbf{k}_\gamma}{|\mathbf{s_n}|\,|\mathbf{k}_\gamma|} + \cdots \qquad (1)$$

An examination of the quantum numbers associated with initial and final states shows that A_γ depends entirely on $\Delta I = 1$ [11, 12].

By a symmetry argument (discussed e.g. in [1]), charged-current $\Delta I = 1$ processes are suppressed relative to neutral-current $\Delta I = 1$ processes by a factor of $tan^2\theta_C$ where θ_C is the Cabibbo angle. Neutral-current processes are therefore expected to dominate $\Delta I = 1$ channels. Neutral-current processes are of particular interest since they do not occur in $\Delta S = 1$ processes. $\Delta S = 0$ processes, and the $\Delta I = 1$ portion in particular, therefore give access to a missing piece of the as-yet incomplete picture of nonleptonic weak interactions.

The pion is expected to play a dominant role in the $\Delta I = 1$ part of the interaction. According to the DDH model, A_γ is expected to depend almost exclusively on h_π^1 [1]:

$$A_\gamma = -0.107h_\pi^1 - 0.001h_\rho^1 + 0.004h_\omega^1, \tag{2}$$

corresponding to a DDH best (most probable) value of $A_\gamma \approx -5 \times 10^{-8}$. The dependence of A_γ on the EFT parameters has been calculated to be [9]:

$$A_\gamma = -0.27\widetilde{C}_6^\pi - 0.09m_N\rho_t, \tag{3}$$

for a six-parameter theory valid at lab energies under 40 MeV, where $\widetilde{C}_6^\pi$ incorporates single-pion exchange and is proportional to h_π^1; and $m_N\rho_t$ is one of the five $S \leftrightarrow P$ amplitudes.

While the PV NN interaction is characterized by total isospin change $\Delta I = 0 \oplus 1 \oplus 2$, in $\Delta S = 1$ decays we have $\Delta I = \frac{1}{2} \oplus \frac{3}{2}$. In the $\Delta S = 1$ case, there is an empirical preference for $\Delta I = \frac{1}{2}$ (known as the $\Delta I = \frac{1}{2}$ rule), but this preference is of unclear dynamical origin. It's possible that an understanding of the isospin dependence of the NN interaction could give us new insight into this problem. Particularly relevant to the $\Delta I = 1$ portion of $\Delta S = 0$, a question has arisen from the interpretation of two measurements: the anapole moment of ^{133}Cs, and the polarization of gamma radiation from ^{18}F decays [3, 13, 14]. The ^{18}F measurement suggests a value for h_π^1 that is at the very low end of the DDH reasonable range. It would be interesting to see if this apparent suppression of a $\Delta I = 1$ weak NN amplitude relative to theoretical expectations is confirmed. The ^{133}Cs measurement on the other hand can be interpreted to give a relatively large value for h_π^1. A model-independent few-body approach may be helpful in addressing the questions arising from this issue.

2. The apparatus

The NPDGamma apparatus was installed at the Los Alamos Neutron Science Center (LANSCE) at Flight Path 12 (FP12) where the first phase of the A_γ measurement was conducted. The apparatus is currently being installed in preparation for the second phase of the measurement at the Fundamental Neutron Physics Beamline (FNPB) at the newly-built Spallation Neutron Source (SNS) at Oak Ridge National Laboratory. Both facilities are pulsed spallation sources, but the FNPB will provide an average neutron rate incident on the target that is about 50 times greater than what was available at FP12.

Figure 1 shows a schematic side view of the NPDGamma apparatus as set up at FP12. The spallation neutrons are produced by 20 Hz, 250 ns long, 800 MeV proton pulses incident on a tungsten target [16] and are backscattered from a 5 cm thick cold hydrogen moderator into the FP12 neutron guide [17]. The average proton current for analyzed data runs was 87 μA. The end of the FP12 neutron guide is located 21 m from the moderator. Due to the high beam flux (a peak instantaneous rate of $\approx 10^{10}$ neutrons/s exiting the FP12 guide) and the impracticality of finely segmenting detectors, the gamma detector array and beam monitors are run in current mode.

The beam monitors are parallel plate ionization chambers containing a mixture of ^{3}He, ^{4}He and N_2 gases and are placed directly in the beam. The cross section for the exothermic reaction

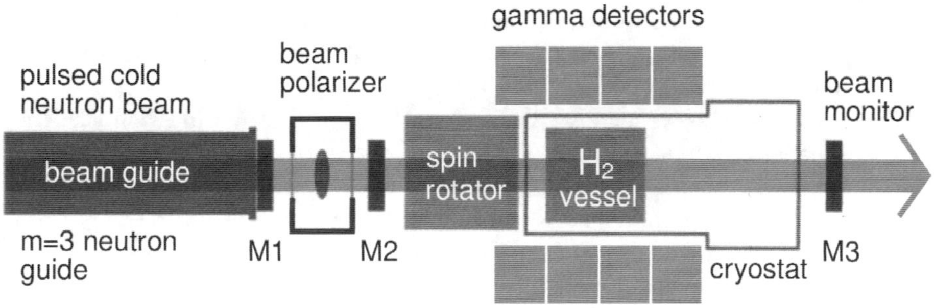

Figure 1. A schematic side view of the NPDGamma apparatus during hydrogen data taking on FP12. Only two opposite detectors of each ring are shown. Not shown are field coils surrounding the entire apparatus which provide a vertical static magnetic field. M1, M2 and M3 are the three beam monitors. The distance from M1 to M3 is about 1.7 m. About 60% of the beam incident on the target is captured by hydrogen, and most of the remainder is absorbed in ^{6}Li-enriched plastic outside the vessel. Only a small fraction of the beam passes through to M3.

$n+{}^3\mathrm{He}\rightarrow p+{}^3\mathrm{H}$ has a well-known $1/\sqrt{E}$ energy dependence. At FP12 and the FNPB there is a direct relationship between time of flight (TOF) and neutron energy since a beam chopper is used to prevent consecutive pulses from significantly overlapping. For fixed beam monitor properties and TOF, the output current is proportional to the rate of neutrons incident on the monitor. Ratios of monitor signals are used to measure relative changes in the transmission of the beam through the apparatus.

The protons for the $\vec{n} + p \rightarrow d + \gamma$ reaction are provided by a 16 liter cryogenic liquid hydrogen vessel [18]. Aluminum was chosen for the cryostat and vessel material because it has small neutron capture and scattering cross sections. Given the 15 meV energy difference between the para and ortho nuclear spin states of the $^1H_2$ molecule, the beam will depolarize in orthohydrogen but will only depolarize significantly in parahydrogen at energies above 15 meV. The equilibrium fraction of orthohydrogen is low enough at operating temperatures ($\approx$ 17 K) that spin-flip scattering is negligible for neutrons below 15 meV. The ortho-to-para conversion rate is increased by circulating the liquid hydrogen through a catalyst. Using data on the energy dependence of scattering cross-sections for orthohydrogen and parahydrogen molecules, the parahydrogen fraction was measured to be 99.9 ± 0.7 % from observations of the energy-dependent transmission of the beam through the target. A description of this measurement and an updated value of the parahydrogen fraction will be given in an upcoming publication [19]. By selection of the appropriate TOF range, the asymmetry measurement excludes data from neutrons with energies that give rise to significant depolarization of the beam in the target.

The beam was polarized at FP12 by transmission through a cell of polarized ^{3}He [20, 21]. The beam polarization is neutron energy dependent [22] and is extracted from fits to *in situ* beam monitor measurements of the transmission of the beam through the cell. The average beam polarization for analyzed data runs was $53 \pm 3\%$. The beam at the FNPB will be polarized using a supermirror polarizer which provides for an improved beam polarization.

The 2.2 MeV gamma rays are detected using a cylindrical array of 48 CsI(Tl) detectors. There are four rings of twelve detectors each arranged contiguously in a cylindrical fashion (see figures 1 and 2). Each 15 cm $\times$ 15 cm $\times$ 15 cm CsI crystal is coupled to a battery-powered vacuum photodiode connected to a low-noise solid state preamplifier. The vacuum photodiodes and beam monitors use the same preamplifier design which is described in [23]. Vacuum photodiodes were used since they possess small nonlinearities ($< 10^{-4}$) when operated in current mode and their gain is insensitive to changes in magnetic field ($< 10^{-3}$/mT). The average detector photoelectron yield was 1300 photoelectrons per MeV [24].

The entire apparatus exists in a static, vertical ≈ 1 mT magnetic field, with vertical gradients in the beam volume less than 0.1 μT/cm. The beam polarization is reversed pulse-by-pulse with efficiency 98.8 ± 0.5 % using the RF spin rotator [25], a solenoid positioned coaxially with the beam that operates by magnetic spin resonance. The neutron spin direction is rotated in the presence of the static field and RF field of the solenoid. The amplitude of the RF field is ramped as 1/TOF so that neutrons of different energies all rotate by the required π radians while traveling down the fixed length of the RF solenoid.

3. The gamma asymmetry measurement

The gamma asymmetry measurement is made using eight consecutive beam pulses with beam polarization ordered temporally according to the sequence ↑↓↓↑↓↑↑↓. A raw asymmetry A_{raw} is extracted for a given detector pair at each of $\approx 50 \times 0.4$ ms-wide time-of-flight bins between ≈ 2 and 15 meV neutron energy using:

$$A_{raw} = \frac{\Sigma_{\uparrow}[N_U - N_D] - \Sigma_{\downarrow}[N_U - N_D]}{\Sigma_{\uparrow}[N_U + N_D] + \Sigma_{\downarrow}[N_U + N_D]}, \tag{4}$$

where N is the pedestal-corrected detector signal. U and D refer to opposing detectors as shown in figure 2. Each sum is performed over one of the two opposite polarization states in the sequence. This 8-step sequence cancels linear and quadratic time dependences of detector properties. To make best use of the dynamic range of the ADCs, the detector signal is read for each ring as one average and twelve differences from the average.

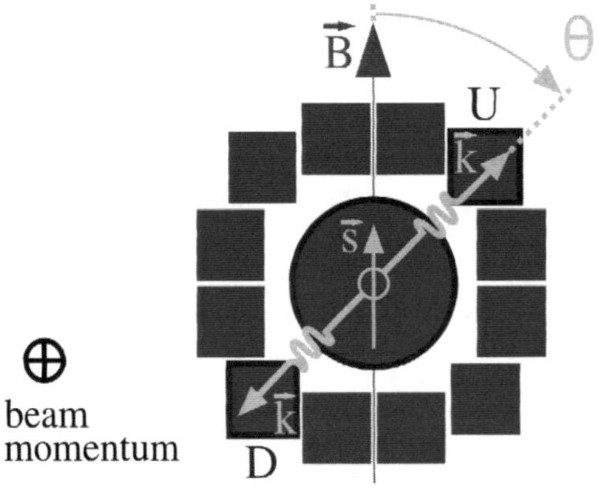

Figure 2. A gamma asymmetry is measured using opposite detectors in a ring which are labeled U and D. U is positioned at angle θ from the direction of beam polarization in the state ↑. $\vec{B}$ is the static magnetic holding field.

A data run typically consists of 1250 8-step sequences. The result presented in this writeup incorporates data from 4966 beam shutter-open, hydrogen target-full data runs. Cuts were made on 1% of the data due to invalid 8-step sequences. At $\approx$ 32-34 ms TOF, the beam chopper closes the beam, allowing for ≈ 6 ms of data that are used to study prompt background (the last 10 ms of the 50 ms pulse are used by acquisition electronics to transfer data). The prompt background was found to be primarily due to activation from neutron capture on aluminum [26]. Separate beam shutter-closed runs were used to observe long-term background and to measure electronic signal offsets (pedestals).

Given the cylindrical symmetry of the detector array, two orthogonal θ-dependent terms are observed simultaneously:

$$A_{raw} = P_n \left(A_{UD} cos\theta + A_{LR} sin\theta \right), \tag{5}$$

where P_n is the beam polarization which has been averaged over the 8-step sequence and corrected for spin-flip efficiency and depolarization ($\approx$ 2% from calculations) in the target. A measurement of A_{UD} or A_{LR} involves averaging over all detector pairs and over many 8-step measurements. Beam intensity fluctuations combined with unequal detector gains contribute to A_{raw} but average to zero with a statistical width an order of magnitude smaller than that of the desired signal.

The appropriate value of θ for each pair was determined from simulations combined with measurements during which the detector array was moved transverse to the beam axis with the target in place. The capacity of the apparatus to observe a θ-dependent γ distribution has been demonstrated by a measurement of the previously-observed large PV asymmetry from polarized neutron capture on ^{35}Cl. Figure 3 demonstrates how this measurement was extracted from 24 individual detector-pair asymmetry measurements. Figure 4 shows a histogram of 8-step A_{UD} measurements that were taken with the parahydrogen target in full production mode.

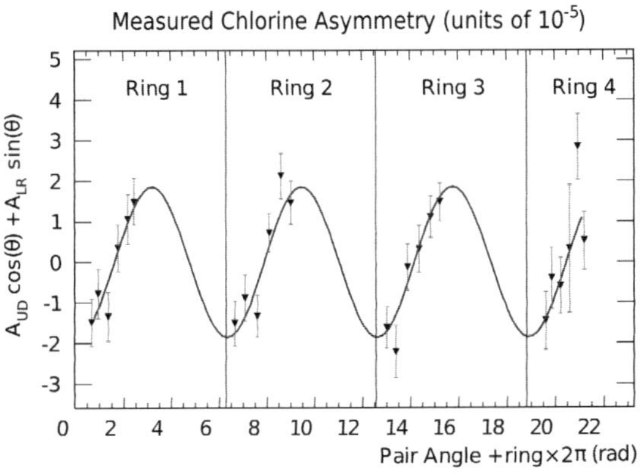

Figure 3. The measured θ dependence of the gamma asymmetry from capture on ^{35}Cl [27]. The $cos\theta$ dependence of the large UD contribution is extracted from a fit while the LR contribution is too small to be seen.

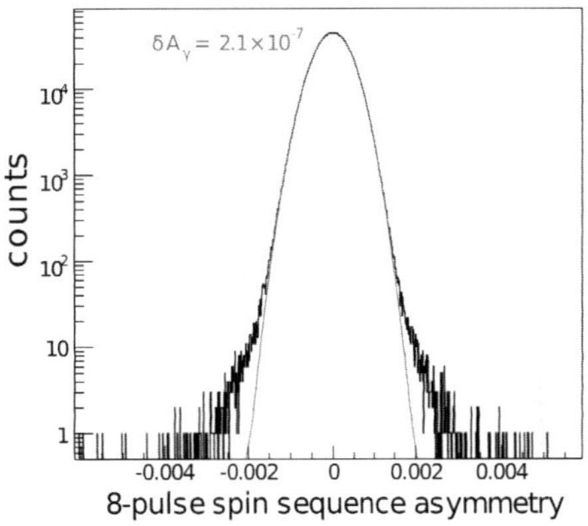

Figure 4. Individual 8-step A_γ measurements histogrammed over many data runs, with a least-squares fit to a Gaussian distribution. The broad wings at low count levels are likely attributable to cosmic ray background.

Effects that mimic A_γ can be separated into beam-on or beam-off effects. Beam-off effects (detector gains and offsets that are correlated with the state of the spin rotator) were measured to be zero at the 10^{-9} level over the period of approximately a day [24]. Beam-on effects are catalogued by considering all cartesian invariants that involve the neutron spin direction and other vectors associated with all of the reaction products that can influence the energy deposition in the detectors. The apparatus is designed to render all such systematic effects from other interactions negligible. The gamma asymmetries from other materials in the apparatus (Al, Cu, In) were measured to confirm the absence of asymmetries from background [28].

Limits on some important beam-on systematic effects are shown in Table 1. The items in the table refer to the mixing of the parity-allowed asymmetry $A_{LR} = (-1.8 \pm 1.9(\text{stat.}) \pm$

0.1(sys.)) $\times 10^{-7}$ into A_{UD} from imperfect detector alignment (good to ± 20 mrad), polarization-dependent γ scattering, up-down steering of the beam due to vertical holding field gradients, and directional asymmetries associated with background processes including neutron decay and capture on materials of the apparatus. Systematic effects are discussed in more detail in an upcoming publication [26]. With systematic effects accounted for, $A_{UD} = A_\gamma$.

Table 1. Limits on important beam-on systematic effects.

Effect	Limit
Left-right asymmetry	1×10^{-8}
γ-ray circular polarization	7×10^{-13}
Stern-Gerlach steering	8×10^{-11}
Beta decay from ^{28}Al	4×10^{-9}
Beta decay in flight	3×10^{-11}
Radiative beta decay	2×10^{-12}
Capture on ^{6}Li	2×10^{-11}

According to a combination of target-empty measurements and MCNPX simulations, a relatively large background contribution of $\approx 25\%$ of the total signal (averaged over all detectors) comes from capture on the aluminum walls and entrance windows of the hydrogen target vessel and cryostat. The asymmetry from aluminum was measured to be consistent with zero with statistical uncertainty at the 2×10^{-7} level. This result was weighted by intensity and subtracted from the hydrogen target-full asymmetry, with statistical uncertainties also appropriately weighted and added in quadrature. Preliminary analysis of the FP12 data [29] indicates an upper limit for A_γ:

$$A_\gamma = (-1.2 \pm 2.1(\text{stat.}) \pm 0.1(\text{sys.})) \times 10^{-7} \tag{6}$$

that matches the precision of the previous measurement $A_\gamma = (0.6 \pm 2.1) \times 10^{-7}$ that was performed in 1977 at the ILL [30, 31]. The goal for the FNPB is to achieve a statistics-limited measurement with uncertainty 1×10^{-8}.

Acknowledgements

The authors would like to thank the US Department of Energy (W-7405-ENG-36), the National Science Foundation (PHY-0100348, PHY-0457219, PHY-0758018 and PHY-0969490), the NSF Major Research Instrumentation program (NSF-0116146) for procurement of the CsI crystals and neutron shielding, the Japanese Grant-in-Aid for Scientific Research (A12304014), the Natural Sciences and Engineering Research Council of Canada, and TRIUMF. The development and application of ^{3}He spin filters was supported in part by the U. S. Department of Energy, Basic Energy Sciences.

References

[1] Adelberger E G and Haxton W C 1985 *Ann. Rev. Nucl. Part. Sci.* **35** 501
[2] Desplanques B 1988 *Phys. Rep.* **297** 1
[3] Ramsey-Musolf M J and Page S A 2006 *Ann. Rev. Nucl. Part. Sci.* **56** 1
[4] Desplanques B, Donoghue J F and Holstein B R 1980 *Ann. Phys.* **124** 449
[5] Desplanques B 1998 *Phys. Rep.* **297** 1
[6] Holstein B R 1981 *Phys. Rev.* D **23** 1618

[7] Zhu S-L, Maekawa C M, Holstein B R, Ramsey-Musolf M J and van Kolck U 2005 *Nucl. Phys.* A **748** 435
[8] Liu C-P 2007 *Phys. Rev* C **75** 065501
[9] Liu C-P 2007 *J. Phys.: Conf. Ser.* **69** 012038
[10] Phillips D R, Schlinder M R, and Springer R P 2009 *Nucl. Phys.* A **822** 1
[11] Danilov G S 1965 *Phys. Lett.* **18** 40
[12] Byrne J 1982 *Rep. Prog. Phys.* **45**, 115 (see pp. 161-2)
[13] Wilburn W S and Bowman J D 1998 *Phys. Rev.* C **57** 3425
[14] Haxton W C, Liu C-P and Ramsey-Musolf M J 2001 *Phys. Rev. Lett.* **86** 5247
[15] Mitchell G S *et al* 2004 *Nucl. Instrum. Meth.* A **521** 468
[16] Lisowski P W, Bowman C D, Russell G J, Wender S A 1990 *Nucl. Sci. Eng.* **106** 208
[17] Seo P-N *et al* 2004 *Nucl. Instrum. Meth.* A **517** 285
[18] Santra S *et al* 2010 *Nucl. Instrum. Meth.* A, in press
[19] Barrón-Palos L *et al* 2010 in an upcoming publication to be submitted to *Nucl. Instrum. Meth.* A
[20] Chupp T E *et al* 2007 *Nucl. Instrum. Meth.* A **574** 500
[21] Gentile T R, Chen W, Jones G, Babcock E, Walker T 2005 *J. Res. Natl. Inst. Stand. Technol.* **110** 299
[22] Rich D R *et al* 2002 *Nucl. Instrum. Meth.* A **481** 431
[23] Wilburn W S, Bowman J D, Penttilä S I and Gericke M T 2005 *Nucl. Instrum. Meth.* A **540** 180
[24] Gericke M T *et al* 2005 *Nucl. Instrum. Meth.* A **540** 328
[25] Seo P-N *et al.* 2008 *Phys. Rev. ST Accel. Beams* **11**, 084701
[26] Gericke M T *et al* 2010 in an upcoming publication to be submitted to *Phys. Rev.* C
[27] Gericke M T *et al* 2006 *Phys. Rev.* C **74** 065503
[28] Gericke M T *et al* 2006 *Phys. Rev.* C **74** 065503
[29] Gericke M *et al* 2006 *Nucl. Instrum. Meth.* A **611** 239
[30] Cavaignac J F, Vignon B, Wilson R 1977 *Phys. Lett.* B **67** 148
[31] Alberi J *et al* 1988 *Can. J. Phys.* **66** 542

XXXIII Symposium on Nuclear Physics IOP Publishing
Journal of Physics: Conference Series **239** (2010) 012013 doi:10.1088/1742-6596/239/1/012013

A measurement of correlation parameters in the decay of polarized free neutrons: the *abBA* experiment

L Barrón-Palos[1], E Chávez[1], C Crawford[2], Q Curiel-García[1], A Huerta[1], M A Juárez-Rosete[1], D J Marín-Lámbarri[1], E Martin[2], M E Ortiz[1], S I Penttilä[3], P Rodríguez-Zamora[1,⋆], A Salas[4], Z Tang[5], W S Wilburn[4]

[1] Universidad Nacional Autónoma de México, México, D.F. 04510, México
[2] University of Kentucky, Lexington, KY 40506, USA
[3] Oak Ridge National Laboratory, Oak Ridge, TN 37831, USA
[4] Los Alamos National Laboratory, Los Alamos, NM 87545, USA
[5] Indiana University, Bloomington, IN 47405, USA

E-mail: `libertad@fisica.unam.mx`

Abstract. The *abBA* experiment will measure, in the same apparatus, four correlation parameters in the free neutron β-decay: the electron-antineutrino angular correlation (a), the Fierz interference term (b), and the asymmetries, with respect to the neutron spin direction, of the electron (A) and antineutrino (B). The precise determination of these parameters, together with the neutron lifetime, will provide important information about the Standard Model (SM) and will establish constraints for new physics. In this paper we describe the experimental methodology of *abBA* as well as some of the advances that have been done so far.

1. Introduction

The measurements pursued by *abBA* will determine with high accuracy the V_{ud} element of the Cabbibo-Kobayashi-Maskawa (CKM) quark-mixing matrix [1, 2], whose nine matrix elements represent the transition amplitudes for weak processes. The accurate determination of the matrix elements is important for unitarity consistency checks of the matrix (weak universality), as well as to establish constraints to physics beyond the SM.

In the case of the unitarity of the top-row of the CKM matrix, where V_{ud} is the most important element, for years the value of the sum $\sum_i |V_{ui}|^2$ was consistently between 2 and 2.5 σ below unity, until recent re-measurements in K-decay resulted in a significant correction to the previous value of V_{us} [3]. With the new value of V_{us}, unitarity with the first row seems now well established ($\sum_i |V_{ui}|^2 = 0.99990(60)$) [4], however uncertainties related to the total transition energy (Q_{EC}) in the superallowed nuclear β-decays from where the present value of V_{ud} is obtained still remain, and must be resolved before a definitive confirmation of unitarity is possible [3, 4].

⋆ Present address: University of Birmingham, Birmingham, B15 2TT, UK

In the superallowed beta-decays only the vector current is involved and therefore $G_V = G_F V_{ud}$, where G_V is the neutron vector form factor and G_F the Fermi coupling constant, which is well known from μ-decay. The neutron lifetime (τ_n) is also defined by the vector and axial-vector neutron form factors (G_V and G_A), or can as well be expressed in terms of V_{ud} and λ, where λ is the ratio of axial-vector to vector neutron coupling constants (g_A and g_V), which at the low momentum transfers that characterize the neutron β-decay are equivalent to the form factors ($\lambda = g_A/g_V = G_A/G_V$)

$$\tau_n = \frac{K}{G_V^2 + 3G_A^2} = \frac{K}{G_F^2|V_{ud}|^2(1+3\lambda^2)}, \tag{1}$$

where K is a constant [3]. From the expression above it can be noted that measurements of τ_n and λ are required to determine the value of V_{ud}

$$|V_{ud}|^2 = \frac{K}{G_F^2 \tau_n (1+3\lambda^2)}. \tag{2}$$

As noted before, at present the most precise determination of V_{ud} comes from superallowed nuclear β-decays. Its extraction from free neutron β-decay, which would have the advantage of being unencumbered by many-nucleon effects present in other nuclear decays, has however not been as precise due to experimental inconsistencies in λ and τ_n. Neutron studies like *abBA* are attempting to resolve these inconsistencies, potentially making the free neutron β-decay the most accurate way to determine V_{ud}.

There is a number of observables related to the neutron decay. The decay rate of polarized neutrons can be expressed in terms of these observables as follows:

$$\begin{aligned} \frac{d^5W}{dE_e d\Omega_e d\Omega_\nu} &= \frac{G_F^2}{(2\pi)^5}|V_{ud}|^2 p_e E_e (E_o - E_e)^2 \xi \left[1 + a\frac{\vec{p}_e \cdot \vec{p}_\nu}{E_e E_\nu} + b\frac{m_e}{E_e}\right. \\ &\left. + \vec{\sigma}_n \cdot \left(B\frac{\vec{p}_\nu}{E_\nu} + A\frac{\vec{p}_e}{E_e} + D\frac{\vec{p}_e \times \vec{p}_\nu}{E_e E_\nu}\right)\right]. \end{aligned} \tag{3}$$

D is the triple correlation coefficient (projection of $\vec{\sigma}_n$ perpendicular to the decay plane) and is not measured by *abBA*. In the vector minus axial-vector ($V-A$) frame of the SM, the four observables measured in *abBA* depend only in one parameter, λ:

$$\begin{aligned} a &= (1-\lambda^2)/(1+3\lambda^2), & b &= 0, \\ B &= 2\lambda(\lambda-1)/(1+3\lambda^2), & A &= -2\lambda(\lambda+1)/(1+3\lambda^2). \end{aligned} \tag{4}$$

The relationships above manifest a very important aspect of the neutron β-decay: that it is overdetermined. From expression 3, at most three parameters are needed to describe the neutron β-decay in the SM: λ, V_{ud} and ϕ (the relative phase between g_A and g_V; parameter D depends on both, λ and ϕ). However, a larger number of observables (τ_n, a, b, B, A, D, etc.) are accessible. This overdetermination of the problem allows to perform different consistency checks of the SM and a sensitive search for new physics.

Parameters a, B and A have been measured experimentally to a precision of 10^{-3} [3]. Sensitivity estimations of the *abBA* experiment, to be performed at the Spallation Neutron Source (SNS) at the Oak Ridge National Laboratory (ORNL), indicate up to an order of magnitude improvement in these three parameters (10^{-4} precision), and in the case of b, the first measurement ever in neutron decay [5].

2. Experimental methodology

The experiment will be performed at the new Fundamental Neutron Physics Beamline (FnPB) at the SNS. Figure 1 shows a scheme of a side view of a transverse cut of the experiment. A pulsed (60 Hz) neutron beam will be polarized either by the use of a ^{3}He spin filter [6, 7] or by a supermirror polarizer [8]; the collaboration is at present evaluating both possibilities. Magnetic field coils provide a holding field for transport of the polarized neutron beam and an adiabatic RF spin flipper (SF) is used to control systematic effects. Neutrons will then be conducted into the decay region, an ultra-high vacuum 60 cm^3 volume defined entirely by magnetic fields so that neutrons do not interact with any matter. The original design of the experiment considered a symmetric decay spectrometer, at the center of which the decay volume is located, as shown in figure 1. The collaboration has recently worked out the new asymmetric spectrometer design for better control of the main systematic effects [9]. The spectrometer is a 4π detector sensitive to decay electron-proton coincidences.

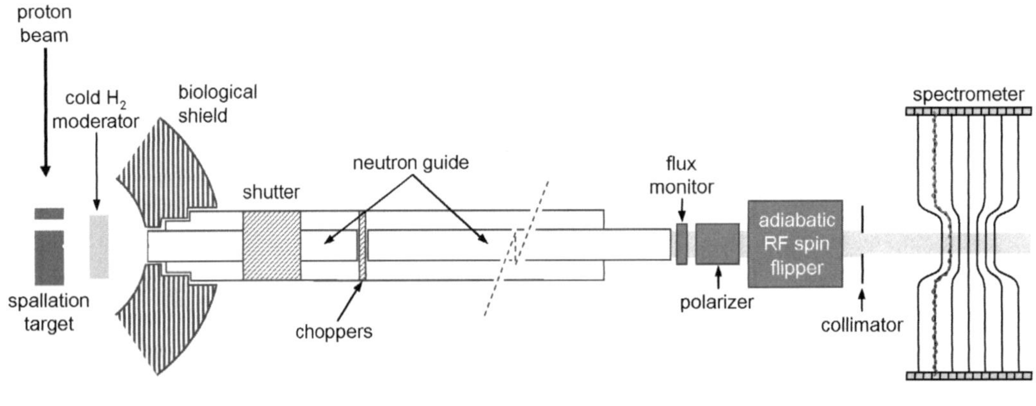

Figure 1. Representation of a side view transverse cut of the *abBA* experiment.

The symmetric spectrometer, conceptually shown in figure 2, consists of cylindrically symmetric magnetic and electric fields whose intensities vary along the spectrometer axis as shown in figure 3, and two segmented silicon detectors at both ends for decay electrons and protons. The intense transverse magnetic field in the decay region (4 T) causes the decay protons and electrons to drift along the magnetic field lines with gyration radii of the order of 1 mm, despite the three orders of magnitude difference in their characteristic emission energies ($E_e >> E_p$). As particles drift from the decay region, they pass through the region where the magnetic field decreases rapidly from 4 T to 1 T. In the frame of the decay particles, the rate of change of the magnetic field is small compared to their gyration frequency, therefore they transport adiabatically. The decrease of the intensity of the magnetic field implies that the gyration radii of the particles increase, reducing their transverse component of momentum while the longitudinal component is increased (adiabatic invariant; conservation of momentum). This longitudinalization of momentum is particularly important for electrons, since it reduces the number of electrons that are Penning-trapped in the high field region and since they hit the detector more normal to its surface, reduces their probability of being backscattered. Additionally, this configuration of the magnetic fields serves as a magnetic mirror that reflects backscattered electrons. Since decay protons are emitted with very low energies (<750 eV), they need to be accelerated for detection. For that purpose, the spectrometer has tubular high voltage electrodes kept at 30 kV in the region of the decay volume and 0 V at both ends of the spectrometer (see figure 3), accelerating the decay protons at 30 keV as they approach the lowest magnetic field region. Since the energy of the decay electrons will be reduced by 30 keV in this electric field, there is a lower limit of 30 keV in the energy of the electrons that can be detected with this apparatus. Electrons and protons are detected in the same 2mm thick silicon detectors, which are about 12 cm diameter circles segmented in 127 hexagonal pixels of about 1 cm^2 of area.

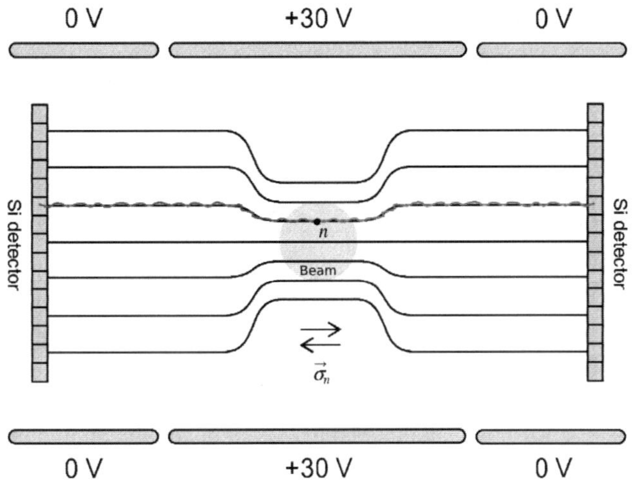

Figure 2. Symmetric 4π neutron decay spectrometer for electron-proton coincidence.

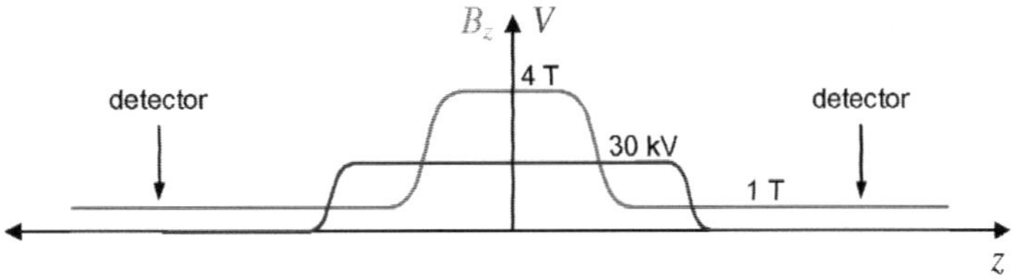

Figure 3. Profiles of electric and magnetic fields along the spectrometer axis.

The design of the spectrometer allows the reduction of backgrounds because the magnetic field lines passing through the decay volume at ultra-high vacuum ensures that the only particles that hit the detectors are those emitted in the decay. Additionally, the spectrometer prevents false coincidences: the decay particles drift along the magnetic field lines with gyration radii that are, in the region of lowest magnetic field, 5mm at most, therefore they are transversally located within two gyration radii (1 cm), a distance which is comparable to the pixel size;

in real coincides, both particles should hit the detector either on the same pixel or in an immediate adjacent pixel (if emitted in the same direction), or in a conjugate pixel in the opposite detector (if emitted in opposite directions). With this apparatus, it will be possible to determine, for each event: the neutron beam polarization, the electron energy, and the direction of the axial component of the electron and proton momenta. This information is used to extract the correlation parameters a, b, B and A. In addition, since decay protons travel first along the magnetic field lines in a region of zero electric field before being accelerated, the time they spend in this region is by far larger than the time they spend in the spectrometer after acceleration. Thus the time difference between the electron and proton arrival provides information of the axial component of the proton momentum, which can be used to obtain information of the kinematics of the decay that improves the sensitivity for the determination of a.

3. Advances

3.1. Si detectors

There is a 20% probability for electrons to be backscattered by the detector, and from those 85% will be reflected back by the magnetic mirror and 15% will escape and hit the opposite detector. In order to reconstruct the latest, which represent 3% of the total, it is necessary to have time resolution of the order of the time of flight of electrons between the two detectors (a few ns), so that it is possible to determine which detector was hit first. Additionally, the detectors must ensure that protons will lose a small amount of energy ($\sim$ 100 eV) in the surface layer. Both requirements set very stringent conditions for the detectors. Fortunately, each of these requirements has been individually met in prototypes and full detector prototypes are at present under testing [10].

3.2. Precision neutron polarimetry

Parameters A and B are directly related to the neutron polarization. To have a 10^{-4} precision in the determination of this parameters, knowledge of the neutron polarization to the 10^{-3} level is required. As mentioned before, one possibility is to use a ^{3}He spin filter to polarize the neutron beam. This is a well established technology that has been extensively used in other experiments, however, the requirements for *abBA* are beyond the present precision limits of the technique, which are typically of a few percent, and therefore development needs to be done. The functioning of these filters is based on the large dependence that the capture of neutrons on ^{3}He has with their relative spin direction. Polarized ^{3}He gas volumes are achieved polarizing the valence electron of a small amount of Rb inside a glass cell (that also contains the ^{3}He) by optical pumping with circularly polarized laser light. Spin exchange with Rb polarizes the ^{3}He. A detailed description of these filters can be found at [6, 7].

The practical difficulties in performing precision neutron polarimetry with this technique can be divided into categories: (i) the physics of the n-^{3}He reaction; (ii) the intrinsic properties of the beam that prevent a perfect relationship between neutron time-of-flight and energy (moderation time, β-delayed neutrons, etc.); (iii) depolarizing interactions with materials and electromagnetic fields that conform the experimental apparatus; (iv) detector backgrounds that affect the neutron transmission measurements from where neutron polarization is determined; and (v) imperfect properties of the ^{3}He spin filters (variations with position in ^{3}He thickness and absorption and scattering of neutrons on the cell glass windows). The uncertainties produced by the first three are below the 10^{-4} level and therefore do not affect neutron polarization at the precision we are seeking. However, the last two represent the greatest difficulties in using this method for precision neutron polarimetry. Backgrounds come from γ-rays or neutrons that are not part of the beam but have scattered into the detector. From these, neutrons are the main concern because the detector efficiency (^{3}He ionization chamber) for gammas is very low compared to that for neutrons. At the Los Alamos Neutron Science Center (LANSCE) of Los Alamos National

Laboratory (LANL) we performed several neutron polarimetry measurements addressing the main identified source of background (neutrons scattered on spin filter glass cell and other beamline components). The analysis of these measurements has not been finalized yet, however significant improvement in precision is anticipated.

3.3. Prototype adiabatic RF spin flipper

To not loose precision in neutron polarization during the spin reversal, a nearly 100% efficient spin flipper is required. This can be achieved by an adiabatic RF spin flipper, which consists of an RF field in the beam propagation direction (z-axis) with a sine modulated amplitude

$$\vec{B}_{RF}(z,t) = A\sin\left(\pi\frac{z+L/2}{L}\right)\sin(\omega_{RF}t)\hat{k}, \tag{5}$$

and a static field along the vertical direction (x-axis) with a cosine modulated amplitude

$$\vec{B}_s = \left[B_o + A\cos\left(\pi\frac{z+L/2}{L}\right)\right]\hat{i}. \tag{6}$$

The spin flipper along the z-axis occupies the region from $-L/2$ to $L/2$ (total length L) and B_o is the amplitude of the vertical static magnetic field used for conserving the neutron beam polarization during transport. Before entering the SF region, the neutron spins precess about this field with Larmor angular frequency given by $\omega_o = \gamma B_o$, where γ is the neutron gyromagnetic ratio. If ω_{RF} matches ω_o, then, in the frame of reference of the neutron spin (a frame rotating about the x-axis with angular frequency ω_o), the effective field produced by the sum of RF and static magnetic fields is constant in amplitude (A) and with direction that changes from being parallel to $\vec{B}_o$ at the beginning of the SF to anti-parallel at the end:

$$\vec{B}_{eff} = \left[A\cos\left(\pi\frac{z+L/2}{L}\right), -A\sin\left(\pi\frac{z+L/2}{L}\right), 0\right]. \tag{7}$$

For neutron spins to continue precessing about the effective field and thus rotate, it is essential that the change of direction of this field is adiabatic; this means that the rate of change of the effective field needs to be small compared to the precession angular frequency of the neutrons about it ($\omega_{eff} = \gamma|\vec{B}_{eff}| = \gamma A$)

$$\frac{\pi}{L/v} << \gamma A. \tag{8}$$

The above condition sets the upper limit in the energy of the neutrons whose spin can be efficiently rotated by 180°. This limit depends on the length (L) and amplitude of the SF fields (A). A first prototype was designed and constructed at the Physics Institute of the National Autonomous University of Mexico (UNAM). The device is shown in figure 4. It consists of a set of four squared coils to produce the static cosine-modulated magnetic field and a solenoid for the horizontal sine-modulated RF field. For the squared coils, if neutron beam propagates from left to right, up and down left coils have the same current and opposite to that of up and down right coils. The solenoid is inside an aluminum can that prevents the RF power to leak out the SF region. The thickness of the can walls is 5 mm, except for the windows on the path of the neutron beam, which are 0.5 mm thick to minimize the neutron absorption/scattering, and yet being thick enough to provide shielding for 29 kHz RF (the Larmor frequency $f_o = \omega_o/2\pi$ for the field $|B_o|$=10 G). In figure 5, the calculated and measured magnetic fields along the SF axis are shown. The total length of the SF is 40 cm and the amplitude of modulation of the component fields is 4 G. With these parameters we get the adiabaticity condition E_n <<456 meV, and the energy of the neutrons used in the experiment is, at most, a few meV. The first prototype testing was performed at LANSCE. Figure 6 shows a frequency distribution of the average neutron transmission through a ^{3}He polarization analyzer located after the SF; two distributions, corresponding to the two spin states can be observed. The efficiency of this prototype has not been measured yet.

Figure 4. The first prototype of the adiabatic RF SF tested at LANSCE.

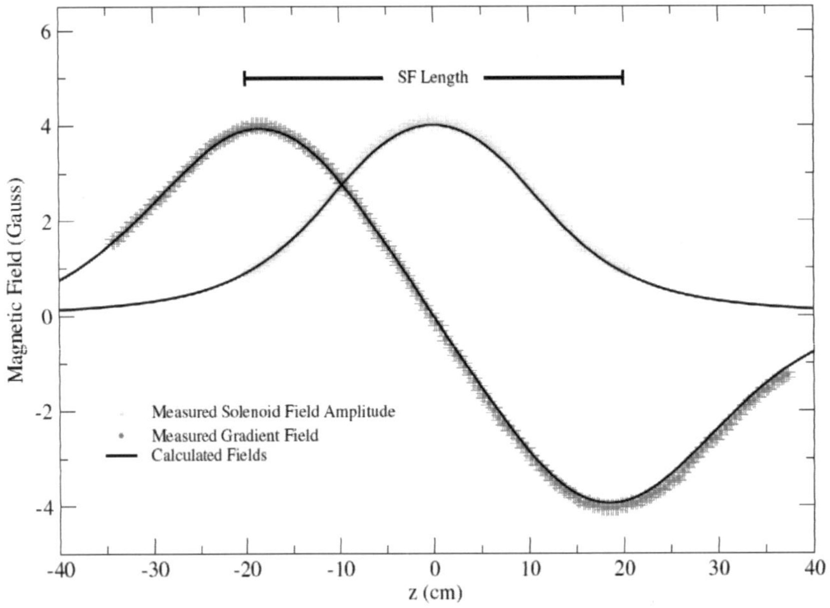

Figure 5. Measured and calculated magnetic field amplitudes along the SF axis.

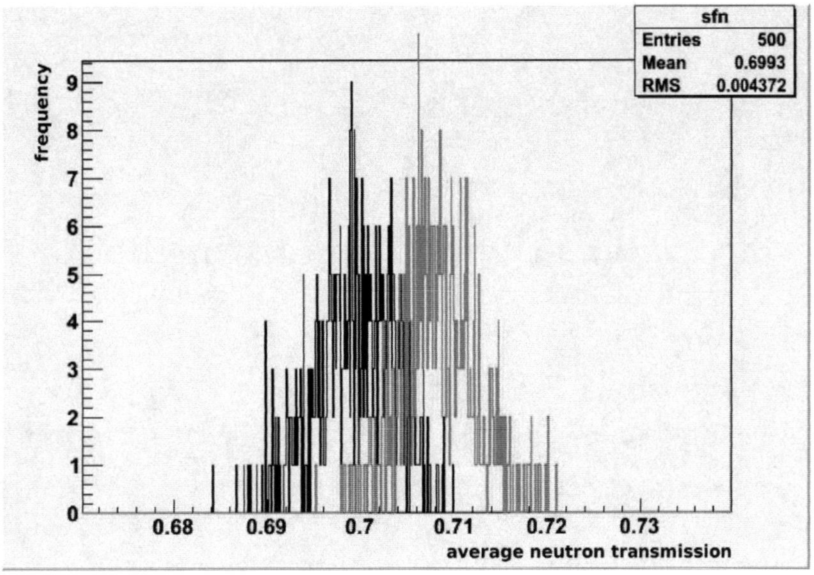

Figure 6. Frequency distribution of the average neutron transmission through a ^{3}He analyzer cell. Two distributions corresponding to the two spin states (SF on and off) can be identified.

4. Conclusions

The *abBA* experiment will be performed within the next few years at the fundamental neutron physics beamline (FnPB) of the ORNL-SNS. Research and development in different aspects of the experiment like neutron beam polarization precision, charged particle detectors with very good time resolution and thin dead-layer, and a highly efficient adiabatic RF spin flipper are currently undergoing. A good degree of advance has been achieved for all of them. Neutron polarization determination to a precision close to 10^{-3}, with ^{3}He spin filters, is expected from the latest measurements at LANSCE. Test measurements with the final version of the charged particle detectors are expected to be performed in the near future. Finally further characterization of the first prototype of the adiabatic RF spin flipper is needed.

Acknowledgments

This work was supported in part by the National Council on Science and Technology of Mexico (CONACYT, contract No. 80444).

References

[1] Cabibbo N 1963 *Phys. Rev. Lett.* **10** 531
[2] Kobayashi M and Maskawa T 1973 *Prog. Theor. Phys.* **49** 652
[3] Amsler C *et al.* (Particle Data Group) 2008 *Phys. Lett.* B **667** 1 and 2009 partial update for the 2010 edition (URL: http://pdg.lbl.gov)
[4] Towner I S and Hardy J C 2010 *Rep. Prog. Phys.* **73** 046301
[5] Wilburn W S *et al* 2007 Precise measurement of neutron decay parameters: the abBA experiment, Beam time request to SNS FNPB Proposal Review Advisory Committee
[6] Chupp T E *et al* 2007 *Nucl. Instrum. Methods Phys. Res.* A **574** 500
[7] Sharma M *et al* 2007 *Phys. Rev. Lett.* **101** 083002
[8] Kreuz M *et al* 2005 *Nucl. Instrum. Methods Phys. Res.* A **547** 583
[9] Baeßler S *et al* 2010 Precise measurement of $\lambda = G_A/G_V$ and search for non-$(V - A)$ weak interaction terms in neutron decay, Funding proposal for the neutron decay spectrometer Nab at SNS
[10] Wilburn W S *et al* 2009 *Rev. Mex. Fís.* **55 (2)** 119

Journal of Physics: Conference Series **239** (2010) 012014 doi:10.1088/1742-6596/239/1/012014

Studies with radioactive ion beams in the $A \sim 80$ region

E Padilla-Rodal

Instituto de Ciencias Nucleares, Universidad Nacional Autónoma de México, Apartado Postal 70-543, 04510 Mexico, D.F., Mexico

E-mail: padilla@nucleares.unam.mx

Abstract. The availability of good quality radioactive ion beams is allowing unique research in nuclear physics and nuclear astrophysics. Coulomb Excitation is a mechanism with a relatively large cross sections that allows the use of low intensity radioactive ion beams, and therefore it has become in a powerful experimental tool for the study of nuclei consirderably far from stability. Among the data of interest for nuclear models that can be obtained using Coulomb excitation are: reduced transition probabilities; electric quadrupole moments; and magnetic moments. This work gives an overview of experimental studies on the mass 80 region in which I have been involved.

1. Introduction

In order to establish the structure of nuclei far from stability it is most important to obtain good-quality data on half-lives, excitation energies, transition probabilities, electromagnetic moments, one- and two-nucleon transfer cross sections, *etc.* The recent availability of Radioactive Ion Beams (RIBs) has made possible to perform pioneering experiments to measure some of these properties [1, 2, 3, 4, 5]. At the Hoilified Radioactive Ion Beam Facility (HRIBF) in Oak Ridge National Laboratory (ORNL), we are actively involved in an experimental program to measure these important quantities in neutron-rich nuclei around mass $A = 80$.

2. Radioactive Ion Beams at HRIBF

HRIBF uses the Isotope Separation On Line (ISOL) technique to produce RIBs [6]. Basically, the $K = 100$ Oak Ridge Isochronous Cyclotron (ORIC) [7] provides an energetic intense light-ion beam (e.g. 10 μA of 50 MeV protons) to induce fission on a thick uranium carbide target. Using high temperatures, $\geq 2000°$C, the fission fragments are diffused out from the target material and flow though a small transport tube that directs them across a high-resistance emission surface. Electrons emitted from this surface are then accelerated into a biased ionization chamber where a plasma is formed and electron impact ionization takes place. The integrated system target/ion source (TIS) is typically operated to $40-60$ kV, and it is the electric potential difference between TIS and the injector platform (typicaly operated to ~ 200 kV) what is used to extract the RIB from the plasma [8].

The injector platform has several beam optics elements and a first mass analyzing magnet (with a mass resolution $M/\Delta M \sim 1000$), and a charge exchange cell that uses re-circulating caesium vapor ($\sim 10^{-2}$ Torr). By collisioning with the Cs gas in the cell, the positive ions coming

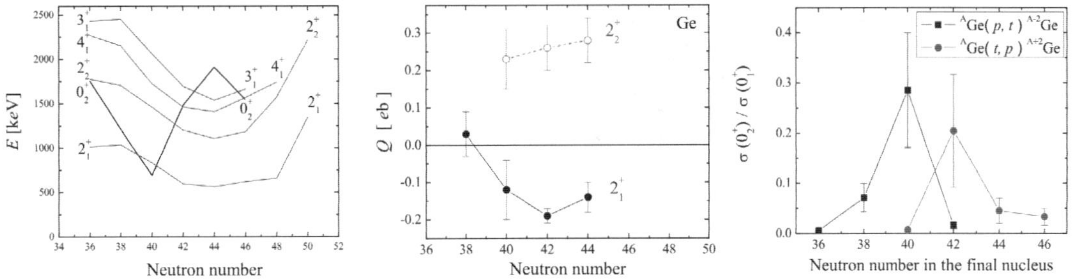

Figure 1. Experimental values for: the excitation energy of the low energy excited states of the germanium isotopes (left); electric quadrupole moments for the states 2_1^+ and 2_2^+ (middle); the population cross section in two nucleon transfer reactions (t, p) and (p, t) (right).

out from the TIS result in a negative beam which is directed into a higher mass analyzing magnet that has a resolving power, $M/\Delta M \sim 20000$. After this second mass selection the ions are finally injected into the 25 MV tandem post-accelerator [9] that supplies the beam energy requested at the experimental station. For the studies that will be described in this contribution, the detection systems placed at the experimental station consisted of a gamma-ray spectrometer operated in coincidence with a charge particle detector and a beam monitor system.

3. The Mass $A \sim 80$ region

Since the mechanism behind RIB production at HRIBF is the induced fission of uranium, it is intuitive that the more intense beams, produced at this laboratory, concentrate around the mass number of uranium's fission fragments: $A \sim 80$ and $A \sim 130$. Our approach in studying exotic nuclei around mass 80, has been trying to get a comprehensive picture of the evolution of its nuclear structure by systematically measuring quantities, such as $E(2^+)$, $B(E2)$, g-factors, quadrupole moments and masses, along isotopic and isotonic chains using almost identical experimental conditions for both stable and radioactive nuclei. Our first step in this approach has been the measurement of the reduced transition probabilities $B(E2; 0^+ \rightarrow 2^+)$ for the neutron rich ^{78}Ge, ^{80}Ge and its $N = 50$ isotope ^{82}Ge [1].

Stable germanium nuclei have a very complex structure, that is reflected in the irregular neutron-dependence of the excitation energy of the 0_2^+ states, the $B(E2)$ values and the population cross sections in two-neutron-transfer reactions [10] (see Figure 1). The abrupt change of these quantities have suggested that this $Z = 32$ chain, undergoes a structural change around $N = 40$. Furthermore, the measured values for the electric quadrupole moments asociated with the 2_1^+ and 2_2^+ states [11, 12], are supporting evidence that a shape transition and the coexistence of two different kinds of deformation are present in these nuclei [13].

A natural question that one might ask is how these properties evolve when more neutrons are added to the nucleus?. The Coulomb excitation (CoulEx) mechanism [14] is a suitable experimental tool that can provide guidance in answering this question. By requesting a bombarding energy smaller than the Coulomb Barrier, one can make sure that the surfaces of the nuclei participating in the reaction do not touch, more precisely, this condition guarantees that the wave lenght of the projectile is smaller than the distance of closest approach in the head on collision, and therefore one can use a semiclassical approach to describe the process, *i.e.* to describe the scattering process clasically, and using the formalism of quantum mechanics to describe the nuclei excitation under the assumption that the only responsable for the excitation is the time dependent electromagnetic field between the two nuclei.

By carefully choosing the bombarding energy one can use CoulEx as a clean tool to measure

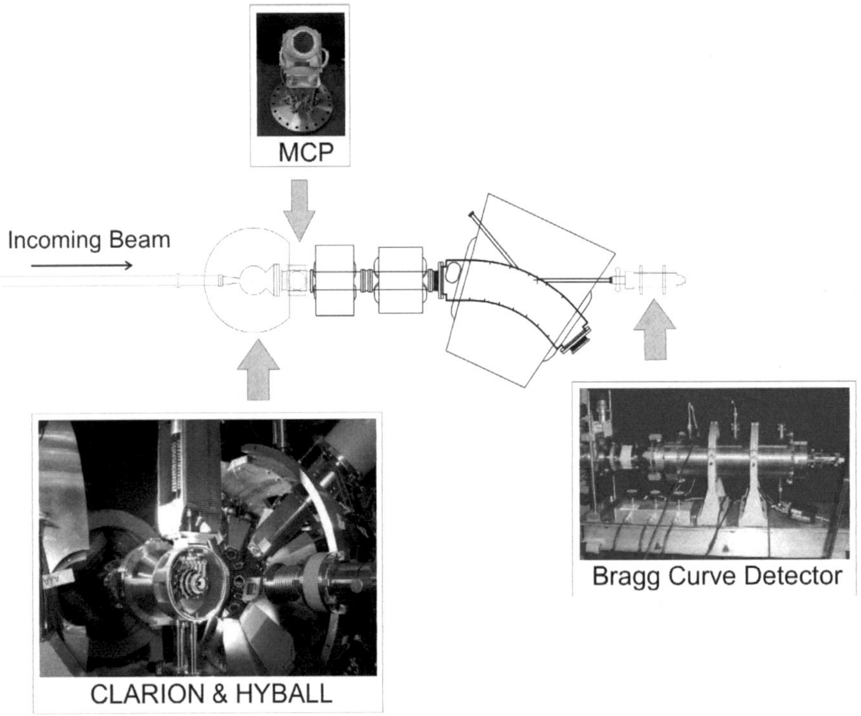

Figure 2. Schematics of one of the experimental station used for nuclear structure studies at HRIBF. The arrows point to the main detector systems shown in the pictures.

excitation energies and transition probabilities of previously unknown nuclei, meaning that one rules out the nuclear interaction and the only responsable for the excitation is the well understood electromagnetic interaction.

4. Detection Systems

Following this idea, we used even-even $A \sim 80$ germanium RIBs produced at HRIBF to bombard light-A targets and study the Coulomb excitation of the projectile nuclei. Detecting the gamma-rays (emmited when the excited nuclei returns to their ground states) in coincidence with the scattering target-like nuclei we were able to measure projectile's lowest energy levels and transition probabilities. Figure 2 shows an aerial view of the detectors at the experimental station, used in the measurement of the $B(E2)$ of ^{80}Ge. The first detector seen by the reaction products is the charge particle array, which is surrounded by a gamma-ray spectrometer. The experimental setup is completed by a beam monitor system located upstream.

Two different technologies in charge particle detectors have been used for CoulEx experiments at HRIBF. One of them is illustrated in Figure 2 by the detector called HyBall [15], which is built of 96 individual elements arranged in 9 rings. Each element consists on a thin CsI(Tl) crystal, optically coupled to a trapezoidal prism of lucite. This geometry is used to guide the scintillation light (generated when a charge particle hit the CsI crystal) towards the readout element consisting on a 1 cm^2 silicon photodiode. The electronic signal is preprocessed using a charge sensitive preamplifier and the final particle identification (PID) is obtained using pulse shape discrimination techniques [16]. For HyBall, the frontal face of each CsI crystal is covered by an absorber foil of lead or tin, and the lateral faces of the lucite prism are wraped up using aluminized mylar (to minimize scintillation light losses). A new version of this array, called BareBall, has been developed more recently with the idea of being used, exculsively, in experiments that involve RIBs. In contrast with HyBall, the angular coverage of BareBall is restricted to forward scattering angles and in order to extend the detection thereshold towards low energies, the frontal face of each element is covered with a very thin layer of aluminized mylar. The second type of charge particle detector consisted on a planar double sided silicon strip detector, S2, designed by Micron Semiconductors [17]. The active region of the detector (35 cm^2) is limmited between an inner diamneter of 22 mm and an outher diameter of 70 mm. The frontal face of this detector is segmented in 48 concentric rings whereas the back face is divided in 16 sectors, providing ΔE, E signals respectively.

For the detection of the de-excitation gamma-rays, we also explored the use of two different kinds of spectrometers. The one shown in Figure 2 is called CLARION [18], and consists on eleven hyper-pure germanium clover detectors. Each detector is formed of four n-type individual crystals segmeted into two halves and sharing a common cryostat. Seven electronic signals can be readout for each clover: four channels of high resolution that are obtained from the central electrode of each crystal (and digitize the energy and time of the gamma-ray detected), and three low resolution channels that come from the external electrode that surounds each half of the crystal (and can be used to get information about the position of the first interaction hit in the detector and also to reduce the Doppler brodening of the gamma-peak). Clover detectors are placed at 25 cm from the target at backward angles to avoid positrons and annihilation radiation background (511 keV). A different spectrometer, operated in coincidence with the double sided silicon strip detector (DSSD) was the so called ORNL-MSU-TAMU barium flouride array, consisting on 150 elements arrenged in four sets of 19 (two of them placed at 90° and two at 135°), and two sets of 37 elements (placed at 89°). Each element is formed by a BaF_2 crystal coupled to twelve-stage, high-gain photomultiplier tubes with a quartz window, wich provides a good transmision of the UV-components of the scintillation light.

An element that has been crucial for the nuclear structure research program at HRIBF, is what we called monitor systems. The basic idea of these ancilliary detectors is to perform a measurement of the isobaric beam compositon by continously sample the beam, after it has interacted with the target. It is worth to emphasize that although desirable, at this point we are not tagging our events using the beam component information, instead we are only getting a rough idea of the isobaric composition by sampling part of the beam. Two ideas have been tested at HRIBF with this purpose: a projectile X-ray system and a Bragg Curve Detector. The projectile X-ray system consisted on a single coaxial HPGe detector placed at 90 degrees relative to the incoming beam direction, which is used to detect the X-rays generated when the beam passed through a thick (~ 4 mg/cm^2) paladium foil. The advantage of this technique is that the isobaric separation does not depend on the energy loss by the projectile, its drawback is however, that the physical process is not well characterized in the range of beam energies that we are interested ($2A - 5A$ MeV). An alternative monitor system is a Bragg Curve Detector (BCD) [19], that consisted on an cylindrical geometry ionization chamber holding an homogeneus electric field in the direction of the incoming particle. The Z identification of the projectile can be done characterizing the amount of ionization (created by the heavy ion in the gas contained inside the chamber), as a function of the projectile position along its slowing-down path. A limitation of a BCD is that it can not hold large count rartes, and it easlily reach saturation if the incoming beam intensity is larger than a few times 10^3 pps.

5. CoulEx experiments

Stable ion beams of the even-even $^{70-76}$Ge, $^{74-82}$Se plus the RIBs of mass $A = 78$ and $A = 80$ were produced and accelerated to $2.4A$ MeV, at HRIBF. These beams were used to bombard a ~ 1 mg/cm^2 carbon target [1].

Typical beam intensities of 1.4×10^6 pps and 1.4×10^5 pps were measured at the target position

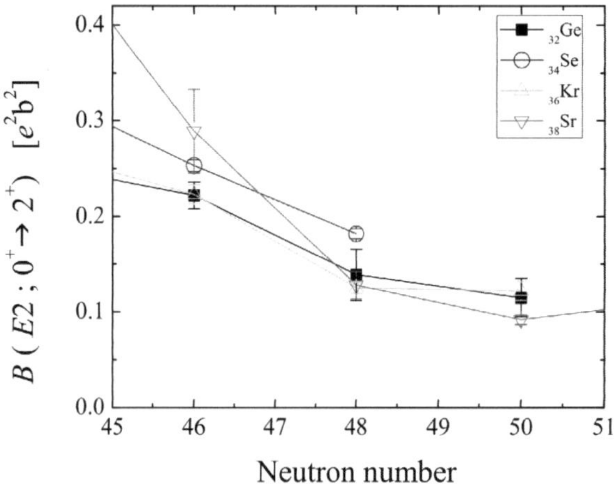

Figure 3. Comparison between the mesured $B(E2)$ values for the radioactive ^{78,80,82}Ge [1] and the trend follow by the adopted values of heavier isotope chains of $_{34}$Se, $_{36}$Kr, and $_{38}$Sr [20]

for RIB $A = 78$ and $A = 80$, respectively. Using projectile X-rays, the beam composition for RIB $A = 78$ was determined to be 57.1% of ^{78}Ge, 28.1% of ^{78}SGe, 9.9% of ^{78}As and 4.9% of ^{78}Ga. For ^{80}Ge, extracting the atomic species from the ion source was not as favorable as it was for ^{78}Ge. Preliminary tests indicated that, only 2.2% of the total X-ray yield was associated with ^{80}Ge, while the ^{80}Se component was 96% and the remanent 1.8% corresponded to ^{80}As, making unfeasable the measurement. It was found that, a good solution to this problem can be acheived by adding sulfur gas (H_2S) to the UC_2 production target, and extracting a molecular beam of GeS^+ (instead of an atomic one: Ge^+) from the ion source. The advantage of doing so was that (because of their chemical properties) the molecular ions of the contaminants: SeS^+ and AsS^+, were suppresed. This effective purification technique, was crucial in the measurement of the $B(E2)$ of ^{80}Ge. After purification, the ^{80}Ge beam component was determined to be 95.3% when measured in the BCD.

The $B(E2)$ measurement of the $N = 50$ nucleus ^{82}Ge was performed using a completly different approach. Based on theoretical predictions it was expected that the reduced transition probability had a lower value compaired with the previously mesure $B(E2)$s. Also since we were moving farther from stability an unavoidable decrease on the beam intensity was expected, and a non-favorable chemistry turned out on a non-significant improve in purity using the sulfur molecule. Given the fact that the excitation energy of the first 2^+ state of ^{82}Ge and its more abundant stable contaminant ^{82}Se can be clearly separated, for this measurement we decided to change the good resolution of the HPGe-based gamma-spectrometer by the larger efficiency of the BaF_2 array, and in order to increase the CoulEx cross section we decided to replace the light carbon target for a heavier ^{48}Ti target, whic also has a well determined $B(E2)$ value.

Figure 3 present the experimental mesurement of the $B(E2)$ values for the radioactive ^{78,80,82}Ge isotopes compaired with the adopted values for the heavier isotope chains of $_{34}$Se, $_{36}$Kr, and $_{38}$Sr, it is important to notice that the behavior of germanium isotopes around $N = 50$ appears to be similar to the plateau exhibited by the isotopic chains with higher-Z.

6. Theoretical Models

These experimental results were compaired with the shell-model theory using a recently derived effective interaction [21, 22]. The calculations were performed in a configuration space that included the orbitals $p_{3/2}$, $p_{1/2}$, $f_{5/2}$ and $g_{9/2}$ [1]. The neutron-neutron (nn) part of such interaction was obtained fitting for nickel isotopes with emphasis on the neutron-rich $^{66-74}$Ni, while the proton-proton (pp) part was obtained fitting the $N = 50$ isotones. The $T = 1$ proton-neutron (pn) part was taken to be identical to nn part, while the $T = 0$ component of such interaction was taken as a G-matrix with modified monopole terms fitted to reproduce the known excitation energies in odd-A Cu isotopes. Full space calculations were performed for ^{82}Ge and ^{80}Ge while no proton excitations to the $g_{9/2}$ orbital were allowed for ^{78}Ge. The theoretical results within the framework of the shell model well reproduced the experimental trend observed for the neutron-rich $^{78-82}$Ge, suggesting that $N = 50$ persists as a good magic number for the germanium chain.

To have an idea on how the shape evolves as we approach to $N = 50$ we also studied of the germanium chain using the Interacting Boson Model (IBM-2) with configuration mixing [23]. In this model, each nucleus is thought as formed of two configurations: normal (spherical) and intruder (prolate), based on the states 0_1^+ and 0_2^+. The normal configuration space consists of two proton-bosons, whereas the intruder has four proton-bosons: one boson-hole in the 20-28 shell and three boson-particles in the 28-50 shell. The calculation proceed by describing each configurations with an independent IBM-2 Hamiltonian of the form

$$H = \varepsilon n_d + \kappa Q_\pi \cdot Q_\nu + M_{\pi\nu}\,, \tag{1}$$

where $n_d = \sum_{\mu,\,\rho}(d^\dagger_{\mu\rho} d_{\mu\rho})$ denotes the number operator of d-bosons, Q_ρ represents the quadrupole operator for protons and neutrons

$$Q_\rho = (s^\dagger_\rho \tilde{d}_\rho + d^\dagger_\rho s_\rho)^{(2)} + \chi_\rho (d^\dagger_\rho \tilde{d}_\rho)^{(2)}\,, \tag{2}$$

and $M_{\pi\nu}$ is the Majorana interaction

$$M_{\pi\nu} = \xi_2\ (s^\dagger_\pi\, d^\dagger_\nu - d^\dagger_\pi\, s^\dagger_\nu)^{(2)} \cdot (s_\pi\, \tilde{d}_\nu - \tilde{d}_\pi\, s_\nu)^{(2)} + \sum_{K=1,3} \xi_K\ (d^\dagger_\pi\, d^\dagger_\nu)^{(K)} \cdot (\tilde{d}_\pi\, \tilde{d}_\nu)^{(K)} \tag{3}$$

after diagonalizing each Hamiltonian in its appropriate space, one constructs a new basis by putting together the five lower energy eigenvectors for each configuration. The mixing calculation consist then in the diagonalization of the Hamiltonian:

$$H_{\text{mix}} = \alpha_0 (s^\dagger_\pi s^\dagger_\pi + s_\pi s_\pi) + \alpha_2 (d^\dagger_\pi \times d^\dagger_\pi + \tilde{d}_\pi \times \tilde{d}_\pi)^{(0)}\,. \tag{4}$$

which does not conserve the number of bosons and connects the two configirations. The additional parameter: Δ, which specifies the unperturbed energy required to excite two protons across the closed shell [24]. Using the eigenfunctions of the two separate configurations one forms the matrix elements of H_{mix}. The final wave functions are obtained from .

In Table 1 we present the most important electric quadrupole transitions between the calculated energy levels for the germanium isotopes. The values are compared with the experimental information available in the literature. The $B(E2)$ values and the quadrupole moments were obtained following the definitions

$$B(E2; L \rightarrow L') = \frac{1}{2L+1}\, |\langle L' || T^{(E2)} || L \rangle|^2\,, \tag{5}$$

$$Q(2_i^+) = \left(\frac{32\pi}{175}\right)^{1/2} |\langle 2_i^+ || T^{(E2)} || 2_i^+ \rangle|^2\,, \tag{6}$$

with the electric quadrupole transition operator given by

$$T^{(E2)} = e_2(Q_{\pi 2} + Q_{\nu 2}) + e_4(Q_{\pi 4} + Q_{\nu 4}), \tag{7}$$

being $Q_{\rho j}$, the quadrupole operator defined in equation (2) for the normal ($j = 2$) and intruder ($j = 4$) configurations. Since in this phenomenologic model we have to use the experimental $B(E2; 2_1^+ \rightarrow 0_1^+)$ values to determine the boson effective charges e_2 (following the work of Sambataro and Molnar [25] we fix $e_4 = 2e_2$ for all isotopes) the actual predictions of this model are quantities such as transition probabilities between higher excitation energy levels, electric quadrupole moments and two nucleon transfer reactions.

7. Present and future research in this region

Three natural extentions of these pioneering studies, which actually define our current research efforts in $A \sim 80$ are: i) the measurement of the $B(E2)$ value for the $N = 50$ nucleus ^{84}Se; ii) the direct measurement of the electric quadrupole moment of ^{78}Ge; iii) the measurement of the g-factor for ^{80}Ge using the Recoil in Vacuum (RIV) technique.

Using projectile Coulomb excitation in inverse kinematics, we have measured the $B(E2)$ value of the radioactive nucleus ^{84}Se. With $Z = 34$, ^{84}Se is the $N = 50$ isotone that lies right in the middle between $Z = 28$ and $Z = 40$ and therefore is expected to have maximum sensitivity to constrain the shell model effective interaction. This value, together with our previously measured ^{82}Ge [1], and the recent result on ^{80}Zn from ISOLDE [4], are providing basic experimental information needed for a better understanding of the neutron-rich nuclei around $A \sim 80$.

One of the most important high-order effects in CoulEx, is perhaps the so call reorientation effect [26], which is a second-order process that consists in the reorientation of the nuclear spin of the Coulomb-excited nucleus caused by the electric field of its reaction partner. This effect can be observed as an interference in the $0^+ \rightarrow 2^+$ transition, and is well described considering that in addition to the direct excitation, the 2_1^+ state is excited through itself (*i.e.* the 2_1^+ itself acts as intermediate state). The change in the nuclear spin direction affects the angular distribution of γ-rays, and what is important is that the additional contribution to the excitation probability of the 2_1^+ coming from the reorientation effect, is proportional to the static quadrupole moment of the excited state. This for instance, gives a measurement of the deviation of the nuclear charge distribution from a spherical symmetry. Following this idea, at HRIBF we have used a purified $A = 78$ RIB at a bombarding energy of $2.3A$ MeV to sequentialy excite ^{78}Ge in two thin targets of ^{24}Mg and ^{12}C. The data, currently under analysis, are being used to calculate the ratio of the CoulEx probabilities for the two different targets, making use of the previously known value of the $B(E2; 0^+ \rightarrow 2^+)$ such ratio will give us a measurement of the quadrupole moment of ^{78}Ge.

The measurement of the magnetic moment for a short-lived excited state is based in determining the precession of the magnetic moment by measuring changes in the angular distribution of the de-exciting γ-radiation. In particular, for the RIV technique, the beam ions after Coulomb excitation emerge into the vacuum, with a total electronic angular momentum, $\mathbf{J}$, that has no directional constrictions. As a consequence the total angular momentum of the nucleus, $\mathbf{I}$, starts to precess about the randomly oriented resultant $\mathbf{F} = \mathbf{I} + \mathbf{J}$. The average over all the ions leads to an attenuated angular correlation between the recoil target ion and the de-excitation γ-emission. When used in combination with RIBs the Recoil in Vacuum technique has proved [27] to have advantages compared with the Transient Field technique for measuring the magnitudes of g-factors, as it does not require a specific optimal positioning for the γ-ray detectors (higher efficiency can be achieved) or the use of a thick target (beam radioactivity does not accumulate).

Table 1. A comparison between the experimental and theoretical $B(E2)$ values and quadrupole moments are given for the Ge isotopes from $A = 68$ to $A = 82$. The units of the $B(E2)$ values are given by $10^{-3}\, e^2$ b^2 while for the quadrupole moments one uses $10^{-2}\, e$b.

	^{68}Ge EXP.	^{68}Ge TH.	^{70}Ge EXP.	^{70}Ge TH.	^{72}Ge EXP.	^{72}Ge TH.
$B(E2; 2_1^+ \to 0_1^+)$	29(3)	27.2	36 (4)	35.9	40 (3)	39.0
$B(E2; 2_1^+ \to 0_2^+)$		4.8	13 (3)	16.5	41 (4)	18.4
$B(E2; 2_2^+ \to 2_1^+)$	0.8(3)	4.2	49.7 (189)	68.2	114 (12)	59.4
$B(E2; 4_1^+ \to 2_1^+)$	22.9(30)	41.0	18.9 (34)	68.1	64.1 (71)	80.0
$Q(2_1^+)$		4.6	3 (6)	2.1	-12 (8)	-6.1
$Q(2_2^+)$		-0.3		9.8	23 (8)	-19.3

	^{74}Ge EXP.	^{74}Ge TH.	^{76}Ge EXP.	^{76}Ge TH.	^{78}Ge EXP.	^{78}Ge TH.
$B(E2; 2_1^+ \to 0_1^+)$	60 (3)	62.2	46 (3)	52.2	44 (3)	40.3
$B(E2; 2_1^+ \to 0_2^+)$	< 7.8	3.0	< 2.8	1.3	$.7\left(^{+5}_{-2}\right)$	3.0
$B(E2; 2_2^+ \to 2_1^+)$	99.7 (203)	91.5	74.6 (96)	73.9	$39.6\left(^{+337}_{-139}\right)$	53.2
$B(E2; 4_1^+ \to 2_1^+)$	66.4 (55)	91.8	73 (13)	74.5	> 21.8	57.4
$Q(2_1^+)$	-19 (2)	-15	-14 (4)	-15.3		-18.3
$Q(2_2^+)$	26 (6)	13.0	28 (6)	11.7		11.9

	^{80}Ge EXP.	^{80}Ge TH.	^{82}Ge EXP.	^{82}Ge TH.
$B(E2; 2_1^+ \to 0_1^+)$	28 (5)	27.6	25 (5)	27.6
$B(E2; 2_1^+ \to 0_2^+)$		3.5		3.5
$B(E2; 2_2^+ \to 2_1^+)$		39.2		39.2
$B(E2; 4_1^+ \to 2_1^+)$		39.0		39.0
$Q(2_1^+)$		-13.6		-0.3
$Q(2_2^+)$		5.2		0.2

For an even-even nucleus the g-factor is sensitive to the way in which angular momentum is shared between all of the protons and all of the neutrons. Since the magnetic properties of protons and neutrons are different from each other, an accurate determination of the g-factors for a series of isotopes offers a unique possibility for studying the influence of the proton-neutron degree of freedom in collective states.

For stable germanium isotopes the g-factor of the 2_1^+ states has been measured using the Ion Implantation Perturbed Angular Correlation [28, 29, 30] and the Transient Field Technique [31, 32]. Recently a systematic analysis of the g-factors around $A = 80$ has been carried out [33]. In the left hand side of Figure 4 the experimental values for the g-factors of stable Ge nuclei have

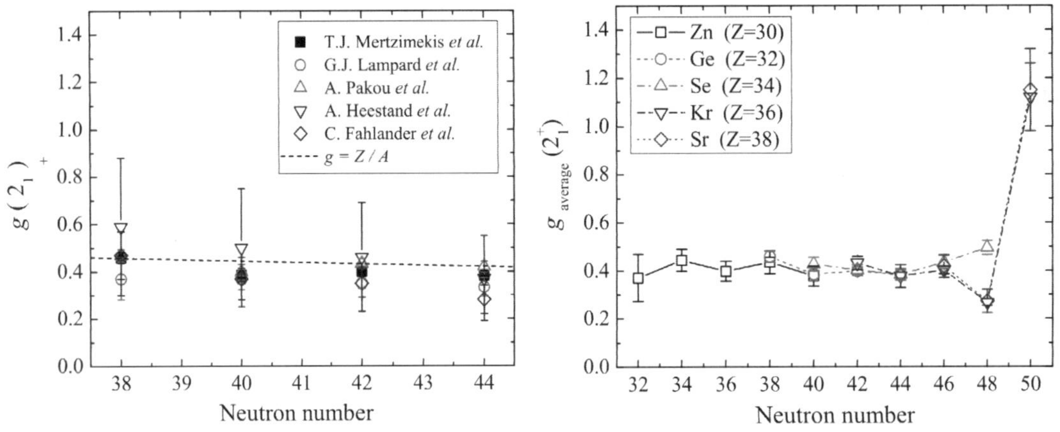

Figure 4. Experimental values for $g(2_1^+)$ of the even-even $^{70-76}$Ge isotopes (left). Systematic behavior of five isotopic chains around $A = 80$ (right).

been plotted , including the average values given in [33]. A very slight variation across the four isotopes is seen, and in general a consistent agreement with the simple hydrodynamic estimate for collective motion, $g = Z/A$ (dashed line) is reached. On the right hand side of Figure 4 the systematic trend for the chains of Zn, Ge, Se, Kr and Sr isotopes is presented as a function of neutron number. It is clear, from this plot, that between $N = 46$ and the shell closure $N = 50$ a dramatic change occurs.

To explore this region in the germanium chain, experimental data using a purified RIB of mass $A = 80$ on a ^{24}Mg target have been taken at HRIBF, to try to measure the g-factor of ^{80}Ge.

8. Conclusions

Since the 60's, CoulEx has been a well known mechanism to study nuclear structure. In recent years, it has received renew interest as a powerful tool to measure transition probabilities, quadrupole moments and magnetic moments of radioactive nuclei. To reach a succesful measurement in combination with RIBs, several challenges associated to the low intensity, isobaric contamination, and intrinsec radioactivity of the beam had to be overcome. However the possibility of using the bombarding energy to select the electromagnetic field as the only responsable for the nucleus excitation makes of CoulEx a valuable and free of ambiguitie tool to study the nucleus. With the second generation of RIB facilities already under construction it is our hope to continue the exploration of more exotic nuclei in region $A \sim 80$ using CoulEx.

Acknowledgments

I would like to thank all my experimental colleagues at HRIBF: A. Galindo-Uribarri, J.P. Urrego-Blanco, D.W. Stacener, J.C. Batchelder, P.E. Mueller, R.L. Vaner, C.-H. Yu, K. Lagergren, D.C. Radford, J.R. Beene and C. Baktash. Special thanks to A.F. Lisetskiy and B.A. Brown for their help with shell model calculations and to O. Castaños and R. Bijker who were involved in the IBM calculations. I also acknowledge the hard work and goodwill from HRIBF operations and beam development staff. ORNL is managed by UT-Battelle, LLC for the U. S. Department of Energy under contract DE-AC05-00OR22725. This research was supported by CONACyT under the grant 103366 and PAPIIT grant IN-121209.

References

[1] Padilla-Rodal E *et al* 2005 *Phys. Rev. Lett.* **94** 122501
[2] Radford D C *et al* 2002 *Phys. Rev. Lett.* **88** 222501
[3] Thomas J S *et al* 2007 *Phys. Rev.* C **76** 044302
[4] Van de Walle *et al* 2007 *Phys. Rev. Lett.* **99** 142501
[5] Stefanescu I *et al* 2008 *Phys. Rev. Lett.* **100** 112502
[6] Beene J R *et al* 2010 to be published in *J. Phys. G: Nucl. Part. Phys.*
[7] Tatum B A 2001 *AIP Conf. Proc.* **600** 148
[8] Carter H K and Stracener D W 2008 *Nucl. Instrum. Methods* B **266** 4702
[9] Jones C M 1981 *Nucl. Instrum. Methods* **184** 145
[10] Vergnes M *et al* 1978 *Phys. Lett.* B **72** 447
[11] Lecomte R *et al* 1980 *Phys. Rev.* C **22** 1530
[12] Toh Y 2002 *AIP Conf. Proc.* **610** 793
[13] Lecomte R *et al* 1982 *Phys. Rev.* C **25** 2812
[14] Alder K and Winther A 1975 *Electromagnetic Excitation* (Amsterdam: North-Holland)
[15] ⟨ http:// www.phy.ornl.gov/hribf/equipment/hyball/ ⟩
[16] Roush M L, Wilson M A and Hornyak W F 1964 *Nucl. Instrum. Methods* **31** 112
[17] ⟨ http:// www.micronsemiconductor.co.uk/pdf/s.pdf ⟩
[18] ⟨ http:// www.phy.ornl.gov/hribf/equipment/clarion/ ⟩
[19] HRIBF NEWS 2002 Edition 10 No. 2 RA3
[20] Raman S *et al* 2001 *At. Data Nucl. Data Tables* **78** 1
[21] Lisetskiy A F *et al* 2004 *Phys. Rev.* C **70** 044314
[22] Lisetskiy A F, Brown B A and Horoi M, 2005 *Eur. Phys. J.* **25** 95
[23] Padilla E *et al* 2006 *Rev. Mex. Fis.* S **52** 57
[24] Van Isacker P *et al* 1986 *Nucl. Phys.* A **451** 202
[25] Sambataro M and Molnar G 1982 *Nucl. Phys.* A **376** 201
[26] De Boer J and Eichler J 1968 *Adv. Nucl. Phys.* vol 1 (Plenum Press, New York) p 1
[27] Stone N J *et al* 2005 *Phys. Rev. Lett.* **94** 192501
[28] Heestand G M *et al* 1969 *Nucl. Phys.* A **133** 310
[29] Fahlander C *et al* 1977 *Nucl. Phys.* A **291** 241
[30] Hubler G K *et al* 1974 *Phys. Rev.* C **9** 1954
[31] Pakou A *et al* 1984 *J. Phys. G: Nucl. Part. Phys.* **10** 1759
[32] Lampard G J *et al* 1987 *Aust. J. Phys.* **40** 117
[33] Mertzimekis T J *et al* 2003 *Phys. Rev.* C **68** 054304

Journal of Physics: Conference Series **239** (2010) 012015 doi:10.1088/1742-6596/239/1/012015

Calculations on the Threshold Anomaly of weakly bound projectiles with São Paulo and Woods-Saxon polarization potentials

A Gómez-Camacho[1], E F Aguilera[1], E Martínez-Quiroz[1], P R S Gomes[2], J Lubian[2] and L F Canto[3]

[1] Departamento de Aceleradores, Instituto Nacional de Investigaciones Nucleares, Apartado Postal 18-1027, C.P. 11801, México, D.F., Mexico

[2] Instituto de Fisica, Universidade Federal Fluminenese, Avenida Litoranea s/n, Gragoatá, Niterói, RJ, cep 24210-340, Brazil

[3] Instituto de Física, Universidade Federal do Rio de Janeiro, C.P. 68528, Rio de Janeiro, R.J., cep 21941-972, Brazil

E-mail: `arturo.gomez@inin.gob.mx`

Abstract. A thorough study of the energy dependence of the nuclear optical potential in reactions involving the weakly bound projectiles ^{8}B, ^{7}Be and ^{6}Li on the target ^{58}Ni and ^{9}Be on ^{27}Al is carried out by performing a χ^2-analysis of recent measurements of elastic scattering cross sections for energies around and above the Coulomb barrier. For this purpose two different potential types are used: the double folding São Paulo potential and the Woods-Saxon potential. The calculations performed for the energy dependence of the real and imaginary parts of the polarization potentials show that these potentials besides satisfying the dispersion relation, for some nuclear systems the uncertainties on the energy dependence of the polarization potentials allow to conclude that these systems present a behavior consistent with the Breakup Theshold Anomaly. In other cases, due to the large uncertainties, it is not possible to make a definitive conclusion about the anomalies.

1. Introduction

Lately, special interest has been focused on the study of the threshold anomaly (TA) at energies around the Coulomb barrier in reactions involving weakly bound projectiles. This interest is due to the fact that the study of the energy dependence of the nuclear optical potential at near barrier energies is one of the tools to investigate the influence of the breakup, nucleon transfer and other reaction mechanisms upon the elastic and fusion channels for reactions with small breakup threshold energies. The TA, usually present in reactions with heavy nuclei, implies a peculiar behavior of the real and imaginary parts of the optical potential around the Coulomb barrier. Both the real and imaginary parts of the optical potential are nearly energy independent at energies well above the Coulomb barrier, but as the energy decreases towards the barrier, the imaginary part $W(E)$ sharply decreases while the real part $V(E)$ presents a localized peak. This correlated behavior is attributed to a strong coupling between the elastic channel and other reaction mechanisms. As the collision energy is lowered, reaction channels such as inelastic scattering, nucleon or cluster transfer, fusion and fission become effectively less important. On the other hand, reaction mechanisms may produce real polarization potentials of different signs, attractive or repulsive, that dynamically decrease or increase the Coulomb barrier [1, 2]. The net real polarization potential depends on the importance and strength of the different specific polarization potentials. This threshold link between the real and imaginary components of the polarization potential is connected by the dispersion relation which has been shown to be a manifestation of the principle of causality [3, 4, 5]. For the case of tightly bound nuclei, for which the most important channel couplings are inelastic excitations of projectile (target) or cluster (nucleon) transfer, the net real polarization potential results attractive at energies near and below the nominal Coulomb barrier. Consequently, these nuclei show enhancement of the fusion cross section at sub-barrier energies when compared to predictions from one dimensional barrier penetration models.

The situation may be different when weakly bound nuclei are involved, due to the strong coupling between the elastic and the breakup channel [6], even at energies well below the Coulomb barrier. As a consequence, the breakup channel remains open to account for the observed large break-up cross sections that overcome the fusion cross section at this energy regime [7, 8]. Accordingly, part of the absorptive imaginary potential $W(E)$ must account for the non- vanishing breakup process and this potential does not vanish at sub-barrier energies. Recent studies suggest that the usual TA is not longer present when weakly bound nuclei are involved [9, 10]. The reason is that as the energy is lowered towards the barrier, a net repulsive polarization potential is produced by the coupling to the continuum breakup states. Accordingly, the strength of the imaginary polarization potential $W(E)$ keeps increasing or at least does not decrease. This effect has recently been observed for several systems involving weakly bound stable projectiles like ^{6,7}Li and ^{9}Be [11, 12]. The conjugate behavior of the real and imaginary polarization potentials is known as the Breakup Threshold Anomaly (BTA) [11]. The BTA has also been observed in the nuclear optical potential of the halo ^{6}He projectile impinging on heavy targets ^{209}Bi [13] and ^{208}Pb [14] at near barrier energies.

Recently, elastic scattering measurements of ^{8}B, its core ^{7}Be and ^{6}Li on the target ^{58}Ni have been performed at energies near and well below the Coulomb barrier[15]. Regarding the system ^{6}Li + ^{58}Ni, other elastic scattering measurements for energies above the barrier have been carried out by Pfeiffer *et al* [16]. Also, for the ^{6}Li scattering, the recent work by Biswas *et al* [17] suggests dubious conclusions about the presence of the TA. The purpose of the present work is to investigate whether the usual TA or the BTA occurs for this weakly bound projectile ^{6}Li, as well as for the proton halo projectile ^{8}B and its core nucleus ^{7}Be reacting on the medium size target ^{58}Ni. Also, the reaction of the weakly bound nucleus ^{9}Be with the target ^{27}Al will be analyzed by means of the polarization potentials as described in the last paragraph. In this case, the analysis will be performed by using the elastic scattering angular distributions and fusion cross sections of Refs. [18, 19]. In order to make a simultaneous fitting of elastic and fusion data, in the case of the Woods-Saxon polarization potential, the imaginary part is split as $W(E,r) = W_F(E,r) + W_{DR}(E,r)$. Being, W_F responsible for fusion reactions and W_{DR} for direct reactions as it will be described below.

Thus, the basic purpose of the present work is to determine by the use of two independent potentials whether or not the threshold anomaly or breakup threshold anomaly appears. That is, in order to obtain potential independent conclusions about the presence of the TA or BTA, we analyze the energy dependence of the real and imaginary parts of the dynamic polarization potential for energies above and below the barrier by two different types of potentials: the double-folding São Paulo potential (SPP) and the Woods-Saxon potential. The barrier energy V_B for the different nuclear systems is calculated with barrier parameters given by the SPP. The São Paulo potential has been widely studied with success in the last years for a variety of systems and reaction mechanisms, including scattering and fusion of weakly bound nuclei, fusion and quasi-elastic barrier distributions and CDCC calculations [20, 21, 22, 23]. In this model the barrier parameters are obtained by approximating the barrier by a parabola at distances near the maximum of the Coulomb and nuclear potentials. The ECIS94 code [24] has been used to perform the χ^2 analyses with the SPP, whereas the JUPITOR code that has been successfully used for many nuclear systems [25], was adopted in the analyses using Woods-Saxon potential. A very brief theoretical description of these potentials is given in section II. The optical potential parameters adequate for each system and for each energy are extracted by fitting the elastic scattering angular distributions for each reaction. The details of this fitting procedure as well as the calculations will be given in the next sections. Finally, we present a short summary and final conclusions.

2. Theoretical Framework

The Hamiltonian H for the nuclear system is of the form,

$$H_a = T_a + \mathcal{V}_a, \tag{1}$$

where the distorted wave $\chi_a^{(+)}$in the incident elastic channel a satisfies the expression,

$$(T_a + \mathcal{V}_a)\chi_a^{(+)} = E_a\chi_a^{(+)}. \tag{2}$$

The potential $\mathcal{V}_a$ is defined by,

$$\mathcal{V}(r,E) = V_{Coul}(r) - U_N(r,E), \tag{3}$$

where the subindex a has been suppressed. Here, $V_{Coul}(r)$ is the Coulomb potential and $U_N(r,E)$ is the energy dependent nuclear potential which is the sum of a bare potential with a slow energy dependence, plus the dynamic polarization potential which may have rapid variations at near barrier energies.

For the nuclear potential, different approaches can be used. As for the bare potential the best physical approach might be to use a double-folding potential for the real part with realistic matter densities and any reasonable approach for the imaginary part. This imaginary part can be constructed with the same radial dependence as the real part (double-folding potential) with a strength coefficient obtained at energies well above the barrier. However, when one studies the optical potential at near barrier energies, strength coefficients are needed to account for dynamic polarizations. A normalized version of the São Paulo optical potential [26, 27] reads as,

$$U_{SP}(r,E) = [N_R(E) + iN_I(E)]F(r,E), \tag{4}$$

where $F(r,E)$ is the double-folding potential

$$F(r,E) = V_F(r)e^{-4v^2/c^2}. \tag{5}$$

Here c is the speed of light, v the relative velocity of the colliding ions given by

$$v^2(r,E) = \frac{2}{\mu}\left[E - V_{Coul}(r) - U_N(r,E)\right], \tag{6}$$

and $V_F(r)$ is the folding potential given in terms of the matter densities of the ions

$$V_F(r) = \int \rho(\mathbf{r_1})\rho(\mathbf{r_2})V_0\delta(\mathbf{r} - \mathbf{r_1}+\mathbf{r_2})d\mathbf{r_1}\mathbf{dr_2}. \tag{7}$$

A zero-range approximation is assumed with strength $V_0 = -456$ MeV-fm^3. In order to provide a parameter-free description of the interaction an extensive systematics of nuclear densities has been proposed by L.C. Chamon [27]. It has been found within the two-parameter Fermi distribution approach to describe the densities that the radii of the distributions are given by,

$$R_0 = 1.31A^{1/3} - 0.84, \tag{8}$$

where A is the number of nucleons in the nucleus. The systematics also shows that the matter diffuseness a of the distributions has an average value $a = 0.56$ fm throughout the periodic table. The normalization of the double-folding potential e^{-4v^2/c^2} arising from the Pauli non-locality due to exchange of the nucleons vanishes at near barrier energies, and therefore the São Paulo potential becomes an usual double-folding potential.

The energy dependent normalization parameters $N_R(E)$ and $N_I(E)$ of Eq. (4) take into account the effects of the dynamic polarization potentials (DPP) arising from direct channel couplings. The real and imaginary parts of the DPPs should be related by the dispersion relation [11]

$$N_R(E) = N_{R0} + \Delta N_R(E), \tag{9}$$

while

$$\Delta N_R(E) = \frac{1}{\pi}\mathcal{P}\int_0^\infty \frac{N_I(E')}{E' - E}dE'. \tag{10}$$

N_{R0} is energy independent and is determined at some high enough energy for which the values of $\Delta N_R(E)$ and N_I are known. The energy dependence of the parameters N_R and N_I are calculated by fitting the elastic scattering angular distributions.

In order to obtain model independent conclusions about the threshold anomaly, a combined use of different approaches for determining the energy dependence of the optical potential is desirable. For this reason, we also use Woods-Saxon shape polarization potentials to calculate the energy dependence of such nuclear polarization potential [28, 29]. In this case, the total interaction between projectile and target of Eq. (2) is

$$\mathcal{V}(r, E) = V_{Coul}(r) - V_0(r) - U_N(r, E), \tag{11}$$

where $V_0(r)$ is the energy independent average nuclear potential and $U_N(r, E)$ is the nuclear dynamic polarization potential given by

$$U_N(r, E) = V(r, E) + iW(r, E). \tag{12}$$

The strength of the real polarization potential $V(E)$ is connected to the imaginary part $W(E)$ by the dispersion relation [3, 4, 5]

$$V(E) = V(E_s) + (E - E_s)\frac{1}{\pi}\mathcal{P}\int_0^\infty \frac{W(r, E')}{(E' - E_s)(E' - E)}dE', \tag{13}$$

where $V(E_s)$ is the value of the potential at the reference energy E_s. The energy independent average nuclear potential $V_0(r)$ and the absorption potential $W(r, E)$ are assumed to have volume geometrical forms. The radial dependence of $V(r, E)$ will be assumed to be the same as that for $W(r, E)$ with the same radial and diffuseness parameters. Therefore, the strengths $W(E)$ and $V(E)$ and the corresponding radial and diffuseness parameters will be extracted from the χ^2-analysis of the elastic angular distribution data as it will be shown in the next section.

For the nuclear systems under study, the energy variation of the parameters $N_R(E)$ and $N_I(E)$ of the SPP is thus studied as the collision energy approaches the Coulomb barrier. These parameters are obtained by fitting the experimental elastic scattering angular distributions for each system. On the other hand, the same energy dependence is searched for $V(E)$ and $W(E)$ of the Woods-Saxon potential whose optical potential parameters are also determined by a χ^2-fitting of the data. Conclusions about the presence of the usual Threshold Anomaly or the Breakup Threshold Anomaly can be drawn from the energy dependence of these functions. All of these calculations are presented in detail in the next section.

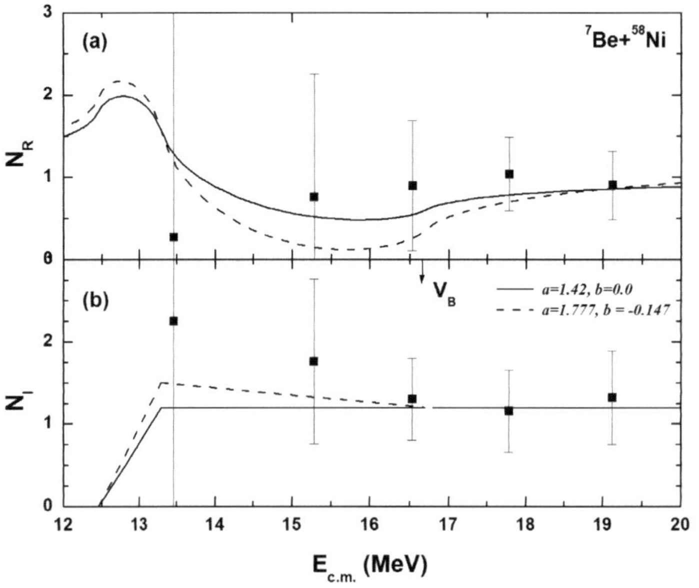

Figure 1. (a) Real N_R parameter of the polarization potential as fitted with the SPP (squares) and the dispersion relation calculation as obtained from the a and b values. (b) Imaginary N_I values as fitted with the SPP (squares) using the data of [15].

3. The Threshold Anomaly Investigation

The recently published data of Aguilera *et al* [15] have been used for the ^{7}Be, ^{6}Li and ^{8}B with ^{58}Ni systems. Fig. 1 shows the values of the N_R and N_I parameters (filled squares) that best fit the elastic experimental data.

The lines in Fig. 1 represent possible behaviors of $N_R(E)$ and $N_I(E)$ that are compatible with the dispersion relation. Different behaviors of N_I can be used in the dispersion relation that follow the "data" obtained from the χ^2-fitting of the elastic angular distributions. The possible linear representations of N_I are given by,

$$N_I = \begin{cases} 0, & E_{c.m.} < E_1 \\ a(E - E_1), & E_1 < E < E_2 \\ a(E_2 - E_1) + b(E - E_2), & E_2 < E < E_3 \\ a(E_2 - E_1) + b(E_3 - E_2) = N_{I\infty} & E > E_3 \end{cases} . \tag{14}$$

The integration of these linear forms have been explicitly given elsewhere [11]. The solid line of Fig. 1a is the result for $N_R(E)$ obtained from Eq. (9) corresponding to the linear fits of Fig. 1b where $E_1 = 12.45$ MeV, $E_2 = 13.4$ MeV and $E_3 = 16.7$ MeV and the a and b values shown in Fig. 1b.

There is a big uncertainty in the determination of the values for E_1, E_2 and E_3. Since, for the nuclear systems under study there are no experimental data in the low energy region to support any extrapolation, thus for this energy region there are several possibilities for the behavior of the imaginary potential. The only thing that we can say is that below a certain energy, the total reaction cross section must vanish and so does the imaginary potential. From the systematic behavior established by Stelson *et al* [30], the quantity, $S = \sqrt{E_{c.m.}\sigma}$ is very close to a linear behavior at low energies, thus it can be represented by $S = \alpha(E - E_{thre})$, α being the slope of

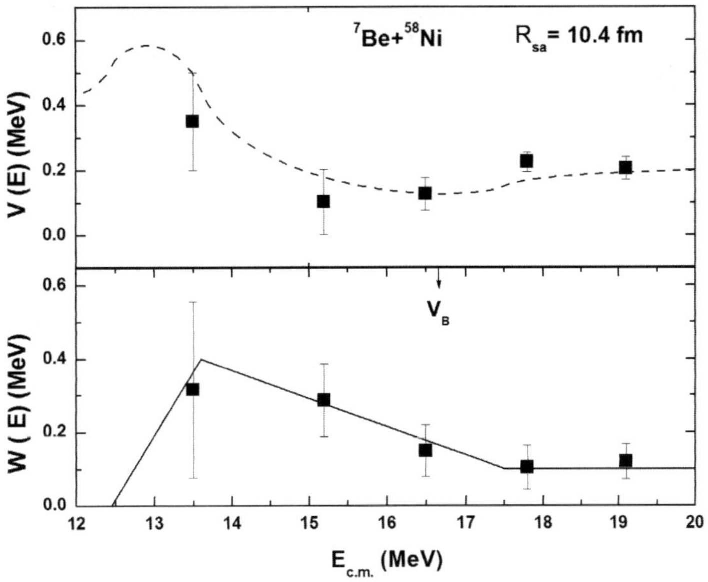

Figure 2. Energy dependence of the Woods-Saxon real and imaginary polarization potentials at the strong absorption radius R_{sa} for the system ^{7}Be + ^{58}Ni. The data for this system correspond to [15].

the linear function and E_{thre} the threshold energy where S or σ become zero. Although, this systematics has been derived from reactions with tightly bound stable nuclei, for the case of weakly bound systems, it can give us an upper limit for the threshold energy. So, the threshold energy E_1 can be determined in this way. On the other hand, E_2 is chosen close to the first data point and E_3 in the region where the values of the imaginary potential N_I seem to stabilize.

The uncertainties of N_R and N_I shown in Fig. 1 have been obtained by calculating the range where these quantities could change by varying the χ^2/N-value in one unity relative to the corresponding minimum value. As observed in Fig. 1b, the error bars of N_I for the lowest energies (up to 3 MeV below the barrier) are too large, not allowing us to draw a definitive conclusion about the presence of the BTA. The reasons for the large uncertainties obtained are possibly due to the poor statistics of the data for this nuclear system, as usually happens when radioactive beams are used, and also to the fact that the SPP has only two normalization parameters with which the data are fitted. The Woods-Saxon potential should give more and the SPP has only two normalization parameters. The Woods-Saxon potential should give more reliable results in such a situation, although both potentials lead to the same conclusions and similar uncertainties when stable projectiles are used [1, 12, 31, 32].

Now, we analyze the same data by using Woods-Saxon potentials. The average nuclear potential used in these calculations is V_0 = 120 MeV, a_0 = 0.56 fm and r_0 = 1.2 fm, these parameters are obtained from the angular distribution at the highest energy. The reduced radius of the imaginary polarization potential has been fixed at 1.4 fm, since the calculated values from the χ^2-analysis become around this value. Therefore, only the strengths of the polarization potential and the diffuseness a_W are fitted. Fig.2 shows the calculations for the polarization potentials $V(E)$ and $W(E)$ at the strong absorption radius R_{sa}.

The dashed line in Fig. 2 is the dispersion relation calculation corresponding to the linear

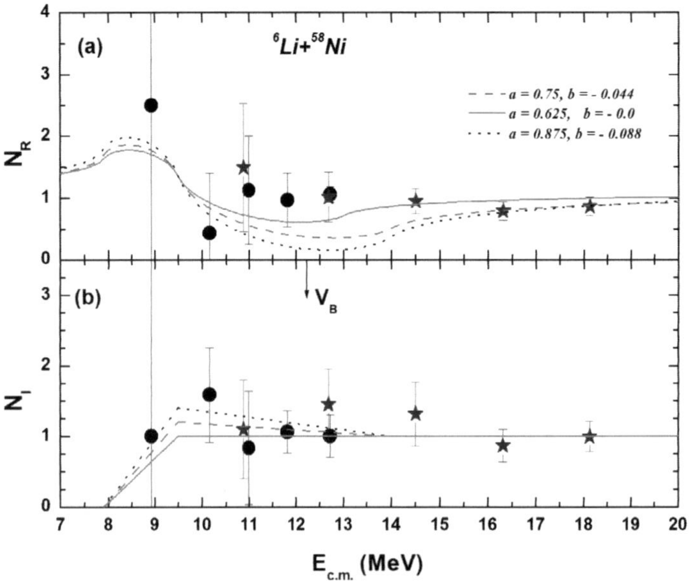

Figure 3. (a) Real N_R parameter of the polarization potential as fitted with the SPP (circles and stars) and the dispersion relation calculation corresponding to the a and b values. (b) Imaginary N_I values as fitted with the SPP (circles and stars). The data represented by circles are taken from [15] and those represented by stars from [16].

fit (full line) of Fig. 2. Here, the reference energy is set at $E_s = 13.6$ MeV corresponding to the reference energy $V(E_s) = 0.45$ MeV. It is observed the repulsive character of the real polarization potential $V(E)$ for energies larger than 13.6 MeV, which in turn raises the Coulomb barrier in this region. On the other hand, the mean values of $W(E)$ show a slight increase as E is lowered below V_B. This increase is still present even for 2 MeV below the barrier. As seen in Fig. 2, the error bars are smaller than in the previous calculation and though there are only three data measurements below the barrier, it is observed that $W(E)$ seems to have a non-decreasing behavior around the barrier. This is a signature that the Breakup Threshold Anomaly shows up for the ^{7}Be + ^{58}Ni system. Here, it is important to emphazise the possible reason of why the calculation with the Woods-Saxon polarization potential gives results with less uncertainties than the one with the São Paulo potential. Besides the fact the elastic scattering angular distributions of [15] report large error bars particularly for the lowest energies, the main possible reason has to be connected to the fact that the Woods-Saxon potential has one more free parameter that can be adjusted to fit the data than the SPP which has only two normalization coefficients.

We perform a similar analysis for the ^{6}Li + ^{58}Ni system. In these calculations, two sets of data have been considered, those of Pfeiffer *et al* [16] and those for lower energies of [15]. So, for this system a thorough analysis is performed, using data at higher and much lower energies than the barrier energy $V_B = 12.3$ MeV. Fig. 3 show the values of the N_R and N_I parameters that best fit the experimental data.

It can be observed that the N_I values obtained from the χ^2-analysis remain rather flat for most of the energies studied. One can also observe that for energies well below the Coulomb barrier N_I does not seem to decrease as it happens when the usual threshold anomaly shows

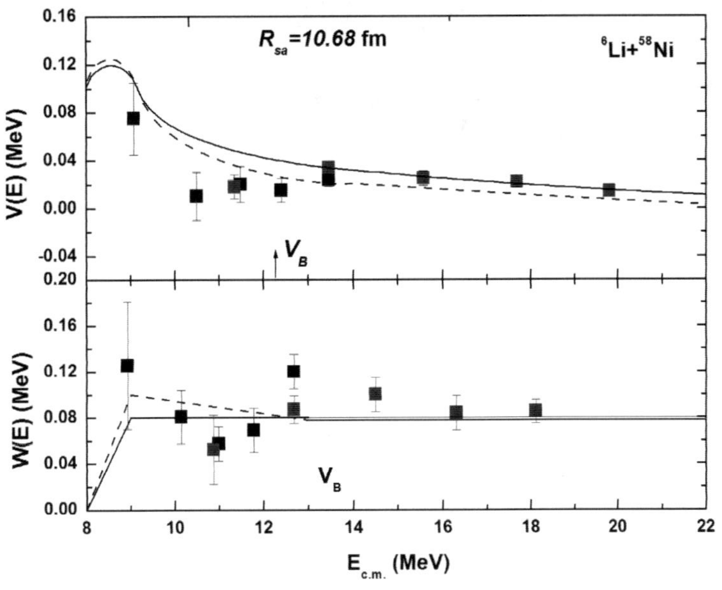

Figure 4. Energy dependence of the Woods-Saxon real and imaginary polarization potentials at the strong absorption radius R_{sa} for the system ^{6}Li + ^{58}Ni. (color online) The data represented with black squares correspond to [15] while those with blue squares to [16].

up. This tells us that an important reaction channel should remain open for energies well below V_B, that is the breakup channel corresponding to the $^6\text{Li} \rightarrow d + \alpha$ channel with small breakup threshold energy of 1.47 MeV. The three lines of Fig. 3a are the results of the dispersion relation integration corresponding to the linear representations of N_I of Eq. (14) with, $E_1 = 7.9$ MeV, $E_2 = 9.5$ MeV, $E_3 = 14.0$ MeV. The anomalous behavior of the imaginary potential at sub-barrier energies corresponds to the presence of the BTA. We would like to mention that although the BTA behavior is usually attributed to break-up, it may be also contain contributions from transfer of nucleons.

As for the previous system, we also performed the analysis using Woods-Saxon potential. An energy independent average nuclear potential $V_0 = 102$ MeV, $r_0 = 1.3$ fm and $a_0 = 0.513$ fm has been obtained by fitting the elastic scattering angular distribution at the highest energy. As for the imaginary potential W, we set r_W at 1.3 fm, thus, only the strengths V, W and the diffuseness a_W are fitted in the analysis. Fig. 4 shows the results for the energy dependence of the polarization potentials V and W.

The error bars have, as before, been calculated by requiring χ^2_{min}/N to vary in one unit *i.e.*, $[(\chi^2_{\text{min}}/N) + 1]$. One can observe that $W(E)$ becomes rather flat for most of the energies while $V(E)$ becomes repulsive and follows well the dispersion relation calculation shown by the line in Fig. 4a. In the dispersion relation, we have set the reference energy at $E_s = 9.0$ MeV with reference energy $V(E_s) = 0.1$ MeV. Similar to the analysis using double folding potential (SPP) , $W(E)$ does not decrease around the barrier. Indeed, it maintains an almost constant value for most of the energy interval and has a slightly increase only for the lowest energies. As it keeps non-decreasing values around the barrier, this is a manifestation of the presence of the BTA for this system.

It has been recently shown that for neutron halo projectiles, like ^6He, the use of double

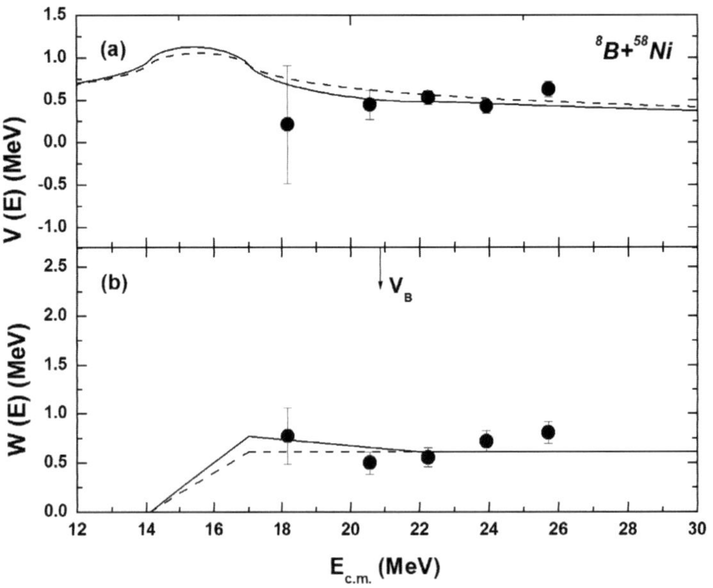

Figure 5. Real (a) and imaginary (b) Woods-Saxon energy dependent strengths of the polarization potential for the system ^{8}B + ^{58}Ni at the strong absorption radius R_{sa}. Elastic scattering angular distributions of [15] have been used in the fitting analysis.

folding potential with strength coefficients for the real and imaginary parts is not a suitable method to simulate the polarization potentials at near barrier energies [33]. Therefore, we do not attempt to investigate the presence of BTA for the proton-halo projectile ^{8}B on ^{58}Ni system using the double-folding potential, but rather only use the Woods-Saxon potential. We calculate $V(r, E)$ and $W(r, E)$ of Eq. (13) at the strong absorption radius R_{sa}. The parameter of the energy independent potential $V_0(r)$ are set as $V_0 = 150$ MeV, $a_0 = 0.56$ fm and $r_0 = 1.2$ fm. These values fit the elastic angular distribution at the highest energy. The parameters of the absorption potential $W(r, E)$ and the strength $V(E)$ are then determined by the χ^2-analysis. In preliminary calculations we have found that the calculated reduced radius r_I becomes around 1.4 fm. So, in the following calculations we have fixed this value. Fig. 5 shows the resulting $V(E)$ and $W(E)$ at the strong absorption radius R_{sa}.

In Fig. 5a, $V(E)$ has been normalized at the values $E_s = 17$ MeV, $V(E_s) = 0.92$ MeV and the lines shown correspond to the dispersion relation integration of the linear fits to the extracted potentials $W(E)$ shown by the full circles in Fig. 5b. Now, for this system ^{8}B + ^{58}Ni, there is only one experimental point at sub-barrier energies to support these calculations, since measurements of elastic scattering at this energy regime are rather difficult and the cross sections show only small deviations from the Rutherford cross section. However from Fig. 5b, it seems clear to us that $W(E)$ does not start to decrease at the barrier energy V_B. On the other hand, the results of the integration of the dispersion relation are compatible with the results obtained from the χ^2-fitting including the lowest energy value for $V(E)$ with large uncertainty. In spite of the fact that there are very few data for energies below the barrier V_B, the results of this calculation show that the conjugated behavior of $V(E)$ and $W(E)$ around V_B indicate that the BTA is present for the system ^{8}B + ^{58}Ni.

The calculations for the system ^{9}Be + ^{27}Al with the São Paulo potential $U_{SP}(r, E)$ are

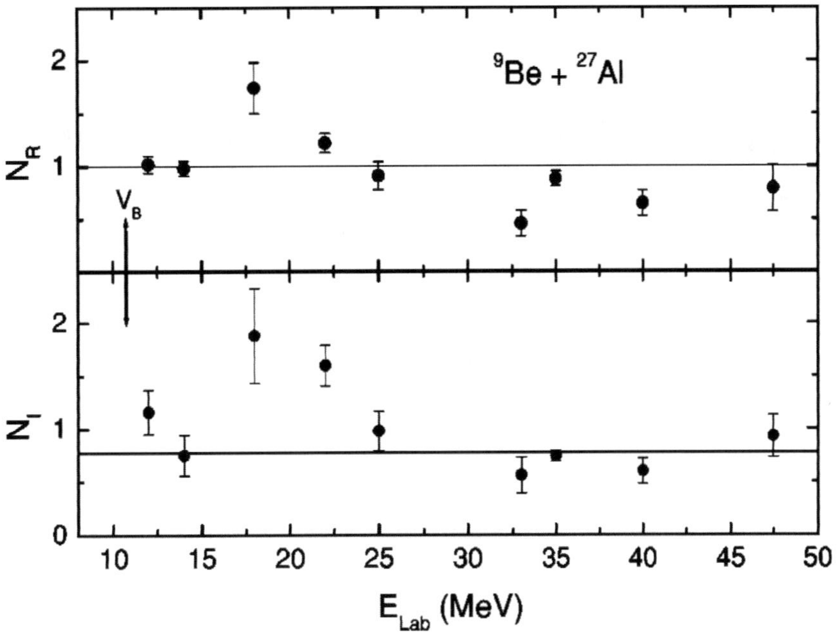

Figure 6. Real N_R and imaginary N_I parameters of the SSP for the reaction ^{9}Be + ^{27}Al. These parameters have been obtained from the best fit to the angular distributions of elastic scattering data of [18].

shown in Fig. 6. Elastic scattering angular distributions are those of [18], where there are not measurements for energies below the barrier. N_R and N_I are the normalization parameters of the real and imaginary parts of the SPP that are determined in the χ^2-fitting to the data. Although the values of N_I show some anomalous increase at two energies near but above the barrier, it is clear that the values of N_I do not decrease as the energy approaches the barrier V_B from high energies. On the other side, as the energy decreases from high energies, the values of N_R show an increasing strength but precisely at the closest energies to V_B this strength decreases. This might be an indication that just around the barrier, the energy variation of N_R manifests a contribution from a repulsive potential that arises from coupling to the breakup channel of ^{9}Be.

The calculations for this system ^{9}Be + ^{27}Al with the Woods-Saxon polarization potentials for the same elastic scattering data are shown in Fig. 7. In this case a simultaneous fitting of elastic scattering with fusion data is performed, where total fusion data are taken from [19]. The polarization potentials $V(r,E)$ and $W(r,E)$ of Eq. (12) are split into fusion parts V_F, W_F and direct reaction parts V_{DR}, W_{DR} respectively. The parameters of these potentials are determined by fitting the fusion σ_F and direct reaction $\sigma_{DR} = \sigma_R - \sigma_F$ cross section data. It is assumed that for this nuclear system, incomplete fusion is negligible as compared to complete fusion in the whole range of energies considered in this work. Also, fusion data at $E_{c.m.} = 13.5$, 18.75, 21.0, 24.0, 24.75, 26.25, 30.0 and 32.2 MeV are those values obtained from a barrier penetration model calculation with parameters $V_B = 8.05$ MeV, $\hbar\omega = 5.1$ MeV and $R_B = 8.39$ fm. The energy independent nuclear potential has parameters $V_0 = 60$ MeV, $r_0 = 1.1$ fm and $a_0 = 0.52$ fm. Regarding the parameters of the imaginary fusion W_F and direct reaction W_{DR} potentials, we set r_F and r_{DR} at 1.4 fm and 1.8 fm respectively. So, the strengths $W_F(E)$, $W_{DR}(E)$ and the diffuseness a_F and a_{DR} are calculated by the χ^2-fitting procedure. As seen in Fig. 7, the values of $V_F(E)$ and $W_F(E)$ show that the fusion potential satisfies the usual threshold anomaly. Also, the values of $W_{DR}(E)$ dominate over $W_F(E)$ at all energies but oscillate very strongly throughout the

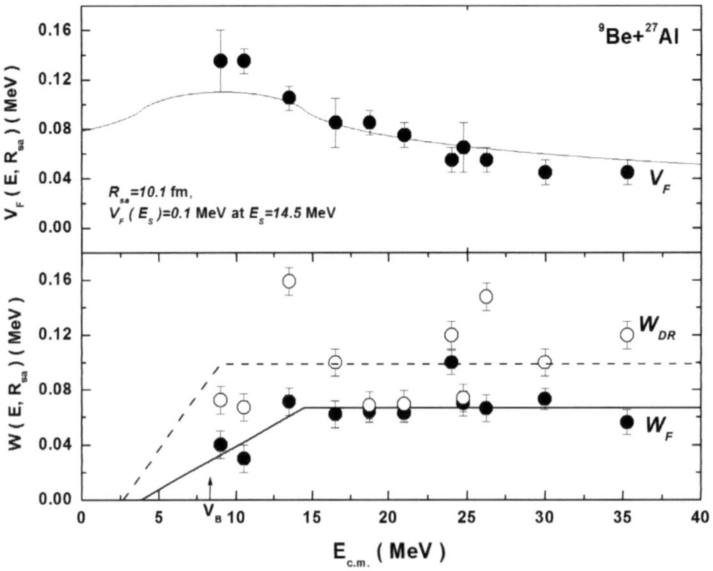

Figure 7. Real and imaginary Woods-Saxon energy dependent strengths of the polarization potential at the strong absorption radius R_{sa} for the ^{9}Be + ^{27}Al system.

energy range of the data. Evidently, the total absorption potential $W(E) = W_F(E) + W_{DR}(E)$ is also affected by these oscillations. From these results, we can not draw any conclusion about the energy behavior of the polarization potentials and therefore about the usual or Breakup Threshold Anomaly. It is possible that the origen of these strong variations on W_{DR} might be related to a) the values used for the fusion cross section at the energies where there are not measurements and that where obtained from a barrier penetration model calculation, or b) that there are some problems with the elastic scattering angular distribution measurements.

4. Summary and conclusions

We have analyzed elastic scattering angular distributions for the weakly bound systems, ^{6}Li, ^{7}Be, ^{8}B + ^{58}Ni at energies around and above the Coulomb barrier. Also, a simultaneous fitting of elastic scattering and fusion cross section data was performed for ^{9}Be + ^{27}Al. The presence of the usual TA, observed for tightly bound systems or the BTA found in some weakly bound systems was investigated through two different approaches, namely, the double folding São Paulo potential and the Woods-Saxon potential. In particular, the energy dependence of the real and imaginary parts of the polarization potentials was calculated around the barrier energy for each system. In this way a potential independent conclusion about the TA or Breakup Threshold Anomaly can be achieved for each system. For some of the systems studied, the determination of the TA or BTA was complicated by the large uncertainties of the real and imaginary energy dependent polarization potentials, particularly when the SPP was used. The calculations of N_R and N_I for ^{7}Be + ^{58}Ni with the SPP show large uncertainties that is not possible to arrive to a definite conclusion about the presence of the BTA. However, the calculation with the Woods-Saxon potential presents less uncertainties so as to conclude that for this system the BTA shows up. Both calculations agree in that the system ^{6}Li + ^{58}Ni shows the BTA. Since the SPP can not be applied to systems with halo projectiles the calculation for ^{8}B + ^{58}Ni was performed only with the Woods-Saxon potential for which the BTA was found. A definite conclusion about the Threshold Anomalies was not possible to make for ^{9}Be + ^{27}Al system with any polarization potential. The difficulties in establishing the presence of the TA or BTA are basically due to the strong variations of the energy dependent polarization potentials N_R and N_I of the SPP as well as the total imaginary part $W(E)$ of the Woods-Saxon potential for energies above the Coulomb barrier. In the latter case, although the direct reaction absorption potential W_{DR} dominates over the fusion potential W_F in the energy region studied. The oscillations of W_{DR} are such that is not possible to establish a certain energy behavior. This is further complicated by the absence of data for energies below the barrier V_B. Certainly, some measurements below the barrier energy and a better statistics above the barrier would allow us to arrive to better conclusions regarding the TA or BTA for the system ^{9}Be + ^{27}Al.

Acknowledgments

Gomes, Lubian and Canto acknowledge the financial support from CNPq, FAPERJ and PRONEX. Gómez-Camacho, Aguilera and Martínez-Quiroz acknowledge the financial support of CONACYT, México.

References

[1] Figueira J M *et al* 2006 *Phys. Rev.* C **73** 054603
[2] Lubian J *et al* 2007 *Nucl. Phys.* A **791** 24
[3] Satchler G R 1991 *Phys. Rep.* **199** 147
[4] Nagarajan M A, Mahaux C C and Satchler G R 1985 *Phys. Rev. Lett.* **54** 1136
[5] Satchler G R and Love W 1979 *Phys. Rep.* **55** 183
[6] Canto L F, Gomes P R S, Donangelo R and Hussein M S 2006 *Phys. Rep.* **424** 1
[7] Gomes P R S *et al* 2004 *Phys. Lett.* B **601** 20
[8] Gomes P R S *et al* 2006 *Phys. Rev.* C **73** 064606
[9] Signorini C *et al* 2000 *Phys. Rev.* C **61** 061603(R)
[10] Gomes P R S *et al* 2005 *J. Phys. G: Nucl. Part. Phys.* **31** S1669
[11] Hussein M S, Gomes P R S, Lubian J and Chamon L C 2006 *Phys. Rev.* C **73** 044610
[12] Figueira J M *et al* 2007 *Phys. Rev.* C **75** 017602
[13] Garcia A R *et al* 2007 *Phys. Rev.* C **76** 067603
[14] Sanchez-Benitez A M *et al* 2008 *Nucl. Phys.* A **803** 30
[15] Aguilera E F *et al* 2009 *Phys. Rev.* C **79** 021601(R)
[16] Pfeiffer K O, Speth E and Bethge K 1973 *Nucl. Phys.* A **206** 545
[17] Biswas M *et al* 2008 *Nucl. Phys.* A **802** 67
[18] Gomes P R S *et al* 2004 *Phys. Rev.* C **70** 054605
[19] Marti G V *et al* 2005 *Phys. Rev.* C **71** 027602
Gomes P R S *et al* *Phys. Rev.* C **70** 054605
[20] Alves J J S *et al* 2005 *Nuc. Phys.* A **748** 59
[21] Gasques L R *et al* 2006 *Nuc. Phys.* A **764** 135
[22] Crema E, Chamon L C and Gomes P R S 2005 *Phys. Rev.* C **72** 034610
[23] Otomar D R *et al* 2009 *Phys. Rev.* C **80** 034614
[24] Raynal J ECIS94 NEA 0850/16
[25] Udagawa T, Tamura T and Kim B T 1989 *Phys. Rev.* C **39** 1840
Udagawa T, Kim B T and Tamura T 1985 *Phys. Rev.* C **32** 124
[26] Chamon L C, Pereira D, Hussein M S, Candido Ribeiro M A and Galetti D 1997 *Phys. Rev. Lett.* **79** 5218
[27] Chamon L C *et al* 2002 *Phys. Rev.* C **66** 014610
[28] Gómez-Camacho A, Aguilera E F and Moro A M 2005 *Nucl. Phys.* A **762** 216
[29] Gómez-Camacho A, Gomes P R S, Lubian J, Aguilera E F and Padrón I 2007 *Phys. Rev.* C **76** 044609
[30] Stelson P H, Kim H J , Beckerman M, Shapira D and Robinson R L 1990 *Phys. Rev.* C **41** 1584
[31] Gomes P R S *et al* 2005 *Phys. Rev.* C **71** 034608
[32] Gómez-Camacho A *et al* 2008 *Phys. Rev.* C **77** 054606
[33] García A R *et al* 2008 *Nucl. Phys.* A **806** 146

AUTHOR INDEX

Adamian, G G 36
Aguilera, E F 100
Alarcon, R....74
Algora, A....36
Antonenko, N V....36
Bahena, D....21
Balascuta, S 74
Barbero, C 59
Barrón-Palos, L....74, 82
Bietenholz, W 66
Bijker, Roelof 53
Bowman, J D 74
Canto, L F....100
Carlini, R D 74
Castro, G López 59
Cejnar, P....7
Chávez, E....82
Chen, W 74
Chupp, T E 74
Cizewski, J A 43
Covrig, S 74
Crawford, C....74, 82
Cseh, J....29, 36
Cundy, N....66
Curiel-García, Q 82
Dabaghyan, M....74
Darai, J 36
Dehnen, H....21
Faleiro, E 1
Fomin, N 74
Fossion, R....7
Frank, A....7
Freedman, S J....74
Gentile, T R 74
Gericke, M T 74
Gillis, R C 74
Göckeler, M....66
Gomes, P R S....100
Gómez, J M....1
Gómez-Camacho, A 100
Greene, G L 74
Hersman, F W 74
Hess, P O....29, 36
Horsley, R 66
Huerta, A 82
Ino, T....74
Jones, G L....74
Jones, K L....43
Juárez-Rosete, M A....82
Klapp, J....21
Kozub, R L 43
Landa, E 7
Lauss, B....74
Lépine-Szily, A 36
Leuschner, M....74
Lévai, G 29
Lozowski, W R....74
Lubian, J 100
Macek, M 7
Mahurin, R 74
Mariano, A 59
Marín-Lámbarri, D J....82
Martin, E....82
Martínez-Quiroz, E 100
Masuda, Y 74
Mei, J....74
Mitchell, G S 74
Molina, R A....1
Muñoz, L 1
Muto, S 74
Nann, H 74
Ortiz, M E 82
Padilla-Rodal, E....90
Page, S A 74
Pain, S D....43
Penttilä, S I....74, 82
Perlt, H 66
Pittel, S 47
Pleiter, D 66
Poves, A 47
Rakow, P E L 66
Ramsay, W D 74
Relaño, A....1
Retamosa, J....1
Rodríguez, L Parra 29
Rodríguez-Zamora, P....82
Salas, A....82
Salas-Bacci, A 74
Sandulescu, N....47
Santopinto, Elena 53
Santra, S....74
Schierholz, G....66
Schiller, A....66
Scholten, Olaf....14
Seo, P-N 74
Sharapov, E I 74
Sharma, M 74
Smith, T B 74
Snow, W M....74
Stránský, P 7
Streuer, T 66
Tang, Z 82
Thakur, B 47
Wilburn, W S....74, 82
Yang, L....47
Yépez-Martínez, H 29
Yuan, V 74
Zanotti, J M 66
Zhao, Y-M....47

Institute of Physics
Dirac House, Temple Back
Bristol BS1 6BE UK

CURRAN ASSOCIATES INC
proceedings.com